Broadening Our Knowledge on Cluster Evolution

T0331444

Although clusters are regarded as important elements in economic development, the strong focus in the literature on the way clusters function is contrasted with a disregard for their evolutionary development: how clusters actually become clusters, how and why they decline and how they shift into new fields and transform over time. Although recently new cluster life cycle approaches emerged, both empirical evidence and theoretical contributions on this topic are still limited. This book therefore contributes to broadening our knowledge on the life cycle and evolution of clusters both empirically and theoretically. It contains chapters on inter-firm relations as drivers of cluster transformation, as well as chapters on the heterogeneity of firms and firm capabilities during cluster evolution and on the role of institutions in stimulating the emergence and growth of clusters. Case studies stem from different industries and technologies, such as biogas, film and television, new media and medical technologies, and from different countries, such as Sweden, Austria, Switzerland and South Korea. All chapters underline that cluster evolution does not only depend on internal dynamics, but that external relations are an integral part of cluster dynamics. This book was previously published as a special issue of *European Planning Studies*.

Dirk Fornahl is Head of the Centre for Regional and Innovation Economics, University of Bremen, Germany.

Robert Hassink is Professor of Economic Geography at Kiel University, Germany, and Visiting Professor in the School of Geography, Politics & Sociology at Newcastle University, UK.

Max-Peter Menzel is Junior Professor of Economic Geography, University of Hamburg, Germany, and Interim Professor of Economic Geography, University of Bayreuth, Germany.

Broadening Our Knowledge on Cluster Evolution

Edited by
Dirk Fornahl, Robert Hassink and Max-Peter Menzel

LONDON AND NEW YORK

First published 2016 by Routledge

2 Park Square, Milton Park, Abingdon, Oxfordshire OX14 4RN
711 Third Avenue, New York, NY 10017

Routledge is an imprint of the Taylor & Francis Group, an informa business

First issued in paperback 2018

British Library Cataloguing in Publication Data
A catalogue record for this book is available from the British Library

ISBN 13: 978-1-138-66616-0 (hbk)
ISBN 13: 978-1-138-39193-2 (pbk)

Typeset in Times New Roman
by RefineCatch Limited, Bungay, Suffolk

Publisher's Note
The publisher accepts responsibility for any inconsistencies that may have
arisen during the conversion of this book from journal articles to book chapters,
namely the possible inclusion of journal terminology.

Disclaimer
Every effort has been made to contact copyright holders for their permission to
reprint material in this book. The publishers would be grateful to hear from any
copyright holder who is not here acknowledged and will undertake to rectify
any errors or omissions in future editions of this book.

Contents

Citation Information

The chapters in this book were originally published in *European Planning Studies*, volume 23, issue 10 (October 2015). When citing this material, please use the original page numbering for each article, as follows:

Chapter 1
Broadening Our Knowledge on Cluster Evolution
Dirk Fornahl, Robert Hassink and Max-Peter Menzel
European Planning Studies, volume 23, issue 10 (October 2015) pp. 1921–1931

Chapter 2
Enterprise- and Industry-Level Drivers of Cluster Evolution and Their Outcomes for Clusters from Developed and Less-Developed Countries
Marta Gancarczyk
European Planning Studies, volume 23, issue 10 (October 2015) pp. 1932–1952

Chapter 3
Born to be Sold: Start-ups as Products and New Territorial Life Cycles of Industrialization
Christian Livi and Hugues Jeannerat
European Planning Studies, volume 23, issue 10 (October 2015) pp. 1953–1974

Chapter 4
Adaptation and Change in Creative Clusters: Findings from Vienna's New Media Sector
Tanja Sinozic and Franz Tödtling
European Planning Studies, volume 23, issue 10 (October 2015) pp. 1975–1992

Chapter 5
Creative Cluster Evolution: The Case of the Film and TV Industries in Seoul, South Korea
Su-Hyun Berg
European Planning Studies, volume 23, issue 10 (October 2015) pp. 1993–2008

Chapter 6
Institutional Context and Cluster Emergence: The Biogas Industry in Southern Sweden
Hanna Martin and Lars Coenen
European Planning Studies, volume 23, issue 10 (October 2015) pp. 2009–2027

Chapter 7

Perspectives on Cluster Evolution: Critical Review and Future Research Issues
Michaela Trippl, Markus Grillitsch, Arne Isaksen and Tanja Sinozic
European Planning Studies, volume 23, issue 10 (October 2015) pp. 2028–2044

For any permission-related enquiries please visit:
http://www.tandfonline.com/page/help/permissions

Notes on Contributors

Su-Hyun Berg worked as a research associate at the Department of Geography, University of Kiel, Germany from 2012 to 2014. She is currently working on her PhD thesis at the Department of Social and Economic Geography at Uppsala University, Sweden. Her research interests focus on the issues of creative industry from an evolutionary economic geography perspective, especially the film and TV industry in South Korea and Sweden.

Lars Coenen is Professor in Innovation Studies at CIRCLE, Lund University and research leader for the platform dealing with sustainability transitions and innovation. He is also a (part-time) research professor at NIFU, the Nordic Institute for Studies in Innovation, Research and Education. His research interests converge around regional innovation systems, eco-innovation and sustainability transitions. Empirically his research is centred on topics related to innovation and the biobased economy as well as sustainable urban transitions. His work has been published in leading international journals such as *Research Policy*, *Environment and Planning A* and the *Journal of Cleaner Production*.

Dirk Fornahl is Head of the Centre for Regional and Innovation Economics, University of Bremen, Germany.

Marta Gancarczyk is Assistant Professor in Institutional Economics and Economic Policy at the Jagiellonian University in Krakow, Poland. Research, publications and consulting focus on public support for small and medium-sized enterprises (SMEs), the growth of the company, managing the competitiveness of SMEs in clusters and networks, strategies for technology companies and the commercialization of innovations.

Markus Grillitsch is a postdoctoral researcher in Economic Geography and Innovation Studies at CIRCLE, Lund University. His main research focuses on the areas of economic geography and innovation studies, specializing in comparative analyses of regional clusters, regional innovation systems, regional innovation policy and differentiated knowledge bases.

Robert Hassink is Professor of Economic Geography at Kiel University, Germany, and Visiting Professor in the School of Geography, Politics & Sociology at Newcastle University, UK.

Arne Isaksen is a Professor at the Institute for Employment and Innovation at the University of Agder. He is also associated with Agder research and has previously worked on STEP-group. His research field is in regional industrial development, with emphasis on studies of regional clusters, innovation systems and regional policy. He has published extensively and internationally on these themes.

Hugues Jeannerat is a postdoctoral researcher at the University of Neuchâtel. His research interests are in the knowledge economy, regional resources and territorial economies, economic sociology of markets and the experience economy.

Christian Livi is a PhD candidate at the University of Neuchâtel. His research currently focuses on territorial innovation approaches, practices and policies. It offers a reflection on regional competitiveness and our way of designing innovation today and for the future. This project aims to establish an interactive and multilateral dialogue between researchers, public officials and various interest groups in the field of economy and regional development.

Hanna Martin is a PhD candidate in Economic Geography at CIRCLE and at the Department of Human Geography, Lund University. Her dissertation deals with the development of cleantech industries in Sweden. In particular, she is interested in studying the regional contexts in which clean technologies emerge and the factors that trigger and hinder their development and diffusion.

Max-Peter Menzel is Junior Professor of Economic Geography, University of Hamburg, Germany, and Interim Professor of Economic Geography, University of Bayreuth, Germany.

Tanja Sinozic is a Lecturer at Vienna University of Economics and Business. She has studied science and technology policy, planning, environmental policy and economics, and is interested in how technologies evolve and interact with the economy and society.

Franz Tödtling is an Associate Professor at the Vienna University of Economics and Business. His research interests are in knowledge-based regional development and innovation systems, networks and clusters, urban and regional development, EU regional policy and globalization.

Michaela Trippl is an Associate Professor in Innovation Studies at CIRCLE, Lund University. Her research and teaching have been dedicated to the fields of economic geography, innovation studies and regional science with a focus on regional clusters, the geography of innovation, long-term regional structural change, labour mobility and regional development, spatial patterns and institutional foundations of the knowledge economy and regional innovation policies.

Broadening Our Knowledge on Cluster Evolution

DIRK FORNAHL*, ROBERT HASSINK** & MAX-PETER MENZEL***

*University of Bremen, Germany, **Kiel University, Germany, ***University of Hamburg, Germany

1. Introduction

The study of regional clusters is a core research topic in economic geography (Markusen, 1996; Martin & Sunley, 2003; Maskell, 2001). Starting from research on industrial districts (Becattini, 2002), the focus has been on the factors that make clusters distinct from spatially dispersed economic activities (Porter, 1998). These studies described the prerequisites of clusters, how particular clusters function at a particular point in time, and how the institutional setting of one place leads to differences in the functioning of clusters at another place (Saxenian, 1994).

Although clusters are regarded as important elements in economic development (Asheim *et al.*, 2006; Martin & Sunley, 2003; Porter, 1998), the strong focus on the way clusters function is contrasted with a disregard for their evolutionary development: how clusters actually become clusters, how and why they decline, and how they shift into new fields and transform over time (Lorenzen, 2005). As a reaction to this gap, and inspired by recent developments in evolutionary economic geography, new life cycle approaches emerged (Staber & Sautter, 2011; Suire & Vicente, 2009; Ter Wal & Boschma, 2011). This strand connects the quantitative development of clusters with underlying qualitative changes and transformations (Bergman, 2008; Menzel & Fornahl, 2010). In doing so, this strand elaborates which dynamics, prerequisites and qualities are connected to the emergence, growth and decline of clusters and in doing so, finds out the patterns of cluster evolution.

The literature mostly differs between the emergence, growth or expansion, decline or transformation of the cluster (Bergman, 2008; Enright, 2003). It has elaborated three different dynamics that drive the cluster through different states of their developments: actors, networks and institutions (Maskell & Malmberg, 2007; Menzel & Fornahl, 2010; Ter Wal & Boschma, 2011). Yet, empirical evidence on these factors from a life cycle perspective is still scarce, and the theoretical contributions describe quite general dynamics, from which single clusters, due to their specific context, most probably deviate.

Furthermore, this strand of literature has a distinct policy perspective. Understanding the dynamics that are connected to the emergence, growth, decline and transformation of clusters allows one to derive targeted policies to govern clusters during different phases of their development (Avnimelech & Teubal, 2008; Brenner & Schlump, 2011). Therefore, this perspective does not solely analyse the path dependencies and contingencies of cluster evolution, but instead the leverages that can be used to contribute to the emergence and growth of clusters as well as to prevent their decline.

This special issue aims to broaden our knowledge on actors, networks and institutions during the life cycle and evolution of clusters in two ways. First, by empirically focusing on hitherto neglected particularities of cluster evolution as well as more nuanced views on actors, networks and institutions in cluster evolution; second, conceptually, by broadening the focus of the prior approaches by integrating different and previously neglected dynamics into this strand of literature. In the following, we first describe new approaches on evolution and life cycles of clusters and how they differ from previous approaches in Section 2. This section also elaborates on actors, networks and institutions as drivers for cluster evolution and through its life cycle. Section 3 describes the contributions of this special issue. We finish this editorial with some thoughts on further research in Section 4.

2. New Approaches on Cyclical Developments

There have always been cyclical approaches to explain the growth and decline of regional industries, for example as a result of the development of new, and the aging of existent, technologies or the change from product to process innovation (Duranton & Puga, 2001; Vernon, 1966). However, these approaches were criticized for being essentialistic, as they consider regional development as a deterministic outcome of often quite general processes (Storper, 1985) and could not explain the different developments of clusters under the same conditions (Saxenian, 1994) or why some industries show another pattern than predicted by the life cycle (Piore & Sabel, 1984).

As a reaction to these critics, and inspired by recent developments in evolutionary economics in general and the evolutionary turn in economic geography in particular (Boschma & Frenken, 2006), new life cycle approaches emerged that focus on micro dynamics instead of the structural dynamics of the old approaches (Menzel & Fornahl, 2010). These new approaches point to three main elements, namely actors, networks and institutions, which differ strongly between different stages of the life cycle and therefore affect the transition from one stage of the life cycle to another. We further explore these three main elements below, but also want to point out here that they are interrelated to each other. Cluster dynamics are in fact derived from the interdependencies of these different elements, something that has been stressed by advocates of the paradigm of relational economic geography (Bathelt & Glückler, 2003; Boggs & Rantisi, 2003). Cluster dynamics are also based on context dependent actions (see also Granovetter, 1985), whereby the context is usually described by institutions or networks. Also Maskell and Malmberg (2007) argue that perceptions of actors that co-evolve with their institutional setting lead to different clustering phases. Ter Wal and Boschma (2011) describe how clusters co-evolve with the respective networks. Menzel and Fornahl (2010) describe knowledge heterogeneity and the way actors utilize it through the life cycle of a cluster.

These new approaches resemble each other in their perspective on clusters not being homogeneous entities. Instead, they derive the dynamics behind cluster evolution from the interplay between heterogeneous agents that change during the cluster life cycle. Due to their focus on micro dynamics, the new approaches overcome the more structuralistic considerations of the older approaches and move from the life cycle analogy towards an ontology of cyclical spatial dynamics (for another view see Martin & Sunley, 2011). In the following we review some literature on the three elements and their interdependencies.

2.1. *Actors*

Single actors, i.e. individuals, firms and public organizations, are emphasized in the cluster life cycle literature because of their role in generating novelty and variety. Research shows the importance of firm-level learning for cluster evolution. Studies such as Klepper's (2007) analysis of the Detroit automobile cluster or Feldman *et al.*'s (2005) description of the Washington biotherapeutics cluster exhibit that firms and their strategies and capabilities are strongly intertwined with the evolution of the cluster. The emergence of technology clusters such as Sophia Antipolis in France (Longhi, 1999) and the Research Triangle Park in North Carolina in the USA (Link & Scott, 2003) also show that political actors can affect the emergence of clusters. Yet, studies such as Grabher's (1993) account on the steel and coal complex in the Ruhr area in Germany also make clear that policy-makers unintentionally can be directly involved in the decline of a cluster.

More recent literature on actors analyses the heterogeneity and capabilities of firms and organizations as well as their learning processes during cluster evolution. Mossig and Schieber (2014), for instance, compare the evolution of a declining and a growing packaging machinery cluster in Germany. They found out that their different development depends on the heterogeneity of firms and determined how firms exploit this heterogeneity. Compared to the declining cluster, the growing cluster has a larger heterogeneity of firms and an institutional environment supporting the exploitation of this heterogeneity. Hervas-Oliver and Albors-Garrigos (2014) demonstrate that during phases of cluster transformation, especially incumbent technological gatekeepers face difficulties dealing with disruptive innovations and that mostly newer firms are able to cope with these changes. But during phases of transformation, large incumbents connect to new firms to secure access to new technologies. Potter and Watts (2014) show that in the declining Sheffield metal cluster Marshallian externalities still prevail, yet are especially important for firms that combine ferrous material knowhow with knowledge form technologically related sectors. In doing so, they point to the importance of Jacobs externalities during phases of restructuring. Elola *et al.* (2012) explain with the example of four clusters in the Basque country in Spain that the initial factor and demand conditions that led to their emergence were no longer important in the later stages. Instead, firms had to build strategic capabilities. Additionally, cluster transformation relied on firm-specific learning processes. Particularly less specialized, more flexible and outward-oriented firms survived phases of cluster transformation. Together, these studies show that heterogeneity and possibilities and capabilities of exploiting heterogeneity are crucial for cluster growth and renewal.

2.2. *Networks*

Studies such as those of Owen-Smith and Powell (2004) on the Boston biotech cluster identify that networks evolve with the evolution of clusters. While firms were connected to public research institutions in the early phase of cluster evolution, they built more firm-centred networks in later phases. On a more abstract level, tight networks are, for example, often connected with the decline of clusters while emerging clusters are shaped by unstable networks (Grabher, 1993; Ter Wal & Boschma, 2011).

Shin and Hassink (2011) analyse the life cycle of the South Korean shipbuilding cluster. They found out that the relations of the growing Korean shipbuilders with the maturity of the cluster became more outward orientated, particularly towards more traditional ship-building clusters such as Norway. Giuliani (2013) shows with the evolution of networks in the Chilean wine industry that networks in developing countries seem to have a different evolutionary pattern. The strong differences in firm capabilities result in quite hierarchical and stable network structures during the evolution of the cluster. Li *et al.*'s (2012) study on the evolution of the aluminium extrusion industry in Dali (China) concludes that the cluster transformed due to political-economic changes and that a new generation of entre-preneurs entered the cluster, thereby changing the network structure from a tight and homogeneous into a more dispersed and formalized structure. These studies found out different forms of contingencies that affect the connection between different phases of cluster evolution and structure of relations. This relation is therefore quite complex. While there seem to be rough generalities in this relation (Ter Wal & Boschma, 2011), context-specific dynamics result in idiosyncratic outcomes.

2.3. *Institutions*

A cluster's institutional setting consists of, for instance, its supportive environments, regional cultures and cognitive frames. Many studies already illustrated how a regional industry culture affects regional development (Piore & Sabel, 1984). Saxenian (1994) and Feldman *et al.* (2005) describe how institutions evolve with the development of clus-ters. Other studies, such as Avnimelech and Teubal (2006), show how institutions, especially state-level institutions, precede the evolution of cluster.

Recent research also stresses the co-evolution between actors and their institutional environment. Staber and Sautter (2011) describe how a local identity evolved with the evolution of the cluster, and different constructions of cluster identity became particularly apparent during cluster upheaval and transformation. In the case study by Tomlinson and Branston (2014) purposive adaption and upgrading resulted in a renewal of the declining ceramics cluster in North Staffordshire in the UK. This renewal also led to the transform-ation of the institutional environment towards new collective supportive organizations, new training and skill courses and collective branding procedures. Skalholt and Thune (2014) compare the reaction of emerging and mature clusters in Norway to the crisis in 2009 and 2010. They discover that especially mature clusters applied strategies of increas-ing collaboration and competence building and that they have developed more capabilities for collective action than newer clusters. These studies agree with accounts on the co-evol-utionary dynamics between cluster firms and their institutional environment (Maskell & Malmberg, 2007). They also point out that their interrelation itself becomes disrupted during phases of cluster transformation, leading to tensions between the established institutional environment and uncertainty about a possible future institutional environment.

All these studies focusing on actors, networks and institutions and their interrelations show that (with the notable exception of cluster decline) a change in quantitative development of the cluster is connected to a qualitative change, marked by changes in firm capabilities, network structures and the institutional environment. They point, in particular, to the disruptions taking place at different levels during cluster transformation.

3. Contribution of the Special Issue

This special issue contributes to this strand of research by adhering to the common framework consisting of the interdependencies between actors, networks and institutions. This framework is used both to connect to the established literature, and to broaden the perspectives on the evolution and life cycle of clusters over different phases. The papers in the special issue are ordered around the dynamics on which they focus. The first two papers focus on actors, in particular on firms and their relations as drivers of cluster transformation.

The theoretical contribution by Gancarczyk (2015) puts the international strategies of lead companies as drivers of cluster evolution into the focus. She argues that contemporary trends of modularization of products and value chains affect the internal organization of clusters as well as the relation between clusters. Additionally, she argues that these trends particularly affect the relation between clusters of developed and developing countries and a more pronounced hierarchical division of labour between them, resulting in R&D in developed, and manufacturing in developing, countries.

Livi and Jeannerat (2015) focus on the relationship between entrepreneurs and multinational companies. Using the case of Swiss medical technologies (Medtech), their paper highlights how the evolution of local Medtech start-ups is shaped by the corporate venture strategies of multinational companies. This strategy allows start-ups to obtain access to capital early. However, this behaviour of multinationals affects start-up processes in a fundamental way. They describe that entrepreneurs starting a firm are less inclined to develop this firm and to produce goods and services in a long-term perspective. Instead, firms themselves are perceived as products that are formed with the intention to be sold. They argue that this change has important implications for the economic development of regions and the life cycle of clusters. They suggest that these finance-led entrepreneurial dynamics create perpetual sequences of the emergence and re-emergence of regional industries.

Both studies have in common that general trends change the evolution clusters. According to Gancarczyk (2015) it is modularization that affects all clusters and might result in a more pronounced hierarchy between clusters. Livi and Jeannerat (2015) demonstrate that the drivers behind cluster evolution change from innovative (i.e. inventing and producing new products that can be sold) towards financial dynamics (i.e. establishing new firms that get access to venture capital funding and can be sold), which also might affect the variety of a cluster. Both studies show that these broad trends result in new patterns of cluster evolution.

While the previous two papers particularly considered inter-firm relations, the next paper by Sinozic and Tödtling (2015) focuses on the heterogeneity of firms and firm capabilities during cluster evolution, using the New Media cluster in Vienna (Austria) as an example. It consists of 480 firms in advertising, video production, information and communication technologies and publishing. Sinozic and Tödtling (2015) indicate that the technological heterogeneity of firms plays a central role in cluster evolution for the expansion of local capacities and opportunities for change, but that it is insufficient as explanation for cluster evolution. Instead, what is required are local technological capabilities, embodied in firms and people, as well as learning conditions for the exploitation of technologies and opportunities.

The next two papers focus on institutions. Berg (2015) describes the emergence and growth of the film and TV cluster in Seoul, South Korea. Around 700 film-making companies are located in Seoul, half of them in the Gangnam Business District. She makes clear how the cluster co-evolved with changes in the broader institutional framework, particularly economic deregulation. She shows that the cluster evolved due to institutional changes on the state level. De-regulation policies allowed other firms to enter the film sector, which led to the growth of the sector and its concentration in Seoul. Currently, the sector is highly affected by location and relocation of important state agencies as part of decentralization strategies of the state as well as by further forms of de-regulations such as the reduction of the yearly screen quota for domestic films. All in all, the case of the film and TV industry in Seoul indicates that its evolution strongly depends on the setting of respective institutions and regulations as well as its institutional surrounding.

In a similar vein, yet with a completely different industry, Martin and Coenen (2015) describe how the institutional context triggered the emergence of the biogas sector in southern Sweden. The cluster consists of firms for feedstock production, collection and transport, pre-treatment and upgrading of biogas, distribution and retail as well as end-use. It consists of about 40 companies that are directly or indirectly involved in the regional biogas value chain. Before the cluster emerged, the prerequisites to form a biogas cluster existed in the region. But it did not emerge before a public programme supported the use of biogas in public transport from 2002 to 2008. After cluster emergence gained momentum, regional institutions also developed to support the cluster.

The case studies by Berg (2015) and Martin and Coenen (2015) therefore describe that clusters did not emerge randomly, but regional prerequisites in connection to state actions, the establishment of new institutions or changes in established institutions affect cluster emergence and transformation by directing resources into or away from a sector. Compared to previous studies, these insights give a more detailed perspective on the co-evolutionary dynamics between clusters and their institutional environment. Institutional changes might precede cluster formation, and institutional changes are the result of strategic and purposeful actions. These outcomes are important for the further development of a policy perspective on cluster evolution, indicating that clusters do not emerge arbitrarily, but that their emergence and evolution can be governed (see also Avnimelech & Teubal, 2006).

The final paper by Trippl *et al.* (2015) adopts a broader, more theoretical approach. They argue for more attention towards the context in which a cluster, its firms and relations, evolve. In this vein, they propose three areas of further research: first, to integrate place specificity and the regional institutional environment of the cluster, second, to consider that networks and institutions that affect clusters are on different scales, each having distinct effects on cluster evolution and third, to consider how human agency and strategic decisions, for example policy actions, have an influence on the evolution of clusters.

While all the papers in this special issue revolve around a broad framework consisting of actors, networks and institutions, some case studies also point to different elements. The papers on inter-firm relations point towards the changing general dynamics of economic organization. The papers on co-evolution between clusters and their institutional environment point towards a sequence where institutional change precedes cluster emergence.

Moreover, all the papers underline that cluster evolution not only depends on internal dynamics. Instead, external relations are an integral part of cluster dynamics. Trippl *et al.* (2015) describe the multi-scalarity of relations and institutions, from local to global scales, as an important research issue in the future. In Livi and Jeannerat (2015) it is the global firm that provides regional firms with capital. In Gancarczyk (2015) global firms' outsourcing strategies result in a stronger hierarchy of clusters. In Sinozic and Tödtling (2015), local firms use the local environment to exploit opportunities and technologies available outside the cluster. In Martin and Coenen (2015), institutional change on the national level provides the framework under which the biogas cluster could emerge. Also Berg (2015) describes how the emergence of the Seoul film and TV cluster started as a reaction towards the change of regulations at the national level.

4. Roads Ahead

This special issue shows the relevance and appropriateness of adopting a life cycle perspective on cluster evolution. Qualitative changes in institutions, networks and heterogeneity of actors are strongly connected to the quantitative change of clusters. More precisely, this special issue confirms the importance of knowledge heterogeneity for the adaptability of clusters. It also concludes that institutional change can precede cluster formation and that the characteristics of relations of cluster firms with firms outside the cluster are not only necessary for cluster renewal, but also affect the actual pattern of cluster evolution. On the basis of these insights three roads for further research are suggested.

First, policy-driven institutional change, such as regulations, affects cluster emergence and growth. In the examples provided in this special issue, investments were directed into the new sector or the state created a new market for a new industry. These institutional effects coupled with fitting regional prerequisites had a strong influence on cluster emergence. While it was common sense that clusters emerged in a kind of accidental and stochastic process (for an early critique see Martin & Sunley, 2006), recent research and also the results of this special issue show otherwise.

Several scholars had already provided insights into how the cluster life cycle concept can inform policies which are better adapted to the stage of the cluster and how these policies differ from static approaches that ignore the life cycle of clusters (Brenner & Schlump, 2011; Hassink & Shin, 2005; Ingstrup & Damgaard, 2013). Examples from Germany allow a comparison of traditional static policies with a life cycle perspective. National cluster programmes, such as BioRegio, InnoRegio or the Spitzencluster-Wettbewerb (Excellence Cluster Competition) as the most recent one (Dohse, 2007; Eickelpasch & Fritsch, 2005), are designed to increase competitiveness and to reinforce existing strengths. This is particularly evident for the Excellence Cluster Competition, which is dedicated to support the most productive and competitive clusters (BMBF, 2009). This kind of strategy bears two kinds of dangers. First, cluster policies might have no additional, or even a negative effect, and this is not recognized, as the future well-being of the cluster is traced back to actual non-effective cluster policies (Fromhold-Eisebith & Eisebith, 2008). Second, this kind of cluster policy reinforces existent cluster structures and developmental patterns, and thus might impede the adjustment of the cluster to changing environments. In doing so it might support its future decline (Hassink, 2010; Menzel & Fornahl, 2010).

In contrast to static approaches, cluster policies that intend to sustain the long-term viability of clusters would focus on supporting the adaptability and changes of a cluster. This not only requires knowledge about the functioning of clusters, but also about the patterns that are connected with their emergence, growth, decline and transformation. Future research would further disentangle the mechanisms and ways of how state actions affect clusters, from regulatory frameworks via supporting institutions to direct involvement.

The second road for further research is the disentangling of what is happening inside and outside a cluster, i.e. the endogenous dynamics of a cluster and when its change depends on external factors. This is apparent both in the papers focusing on institutions and those analysing relations. In both cases, the benefits of adopting a multi-scalar perspectives as proposed by Trippl *et al.* (2015) become obvious. There is already a large body of literature on how relations at different geographical scales affect clusters (Bathelt *et al.*, 2004) or how external factors, such as field configuring events, change a field (Lampel & Meyer, 2008). It would make sense to combine the strand on cluster evolution and life cycles stronger with that on different spatial scales of relations. Of relevance would be which kind of relations serve which kind of purpose in which phase of cluster evolution. This would allow devising more tailor-made policy measures.

The third road of research revolves around the effects of socio-economic and technological changes on cluster evolution. There is already literature that describes how disruptions in the technologies applied by cluster firms result in decline and transformation of established clusters (Dalum *et al.*, 2005). Yet, of further interest would be how general technological or socio-economic change affects the general pattern of cluster evolution. Leamer and Storper (2001) already hypothesized in 2001 that the upcoming of the internet would have an effect on the geography of production and the spatial fragmentation of economic activity in space. Research on clusters was revitalized by industrial district research, which resulted from a shift from Fordist to post-Fordist flexible forms of production (Boyer, 1997). Such a shift might again occur. Financialization (French *et al.*, 2011) might be such a development that fundamentally affects evolutionary patterns of clusters, i.e. how firms utilize local diversity.

This special issue contributes to further considering the complexity of the interrelations between actors, networks and institutions during the evolution and life cycle of clusters. Yet, there are still many unresolved questions, particularly about the contextualization of cluster evolution. Further connecting the dynamics of qualitative change to the quantitative development of clusters is still required for meaningful, theory and fact-based policies regarding cluster development in particular and regional development in general.

Acknowledgments

This special issue draws heavily from papers presented at special sessions on cluster life cycles at the Regional Studies Association Annual International Conference in Tampere, 5–8 May 2013. All authors, except for Gancarczyk (2015), are project partners in the research project 10-ECRP-007: Cluster life cycles—the role of actors, networks and institutions in emerging, growing, declining and renewing clusters, which is sponsored by the European Science Foundation in the framework of the program "European Collaborative Research Projects in the Social Sciences" (ESF-ECRP).

References

Asheim, B., Cooke, P. & Martin, R. (Eds) (2006) *Clusters and Regional Development: Critical Reflections and Explorations* (London: Routledge).

Avnimelech, G. & Teubal, M. (2006) Creating venture capital industries that co-evolve with high tech: Insights from an extended industry life cycle perspective of the Israeli experience, *Research Policy*, 35(10), pp. 1477–1498. doi:10.1016/j.respol.2006.09.017

Avnimelech, G. & Teubal, M. (2008) Evolutionary targeting, *Journal of Evolutionary Economics*, 18(2), pp. 151–166. doi:10.1007/s00191-007-0080-6

Bathelt, H. & Glückler, J. (2003) Toward a relational economic geography, *Journal of Economic Geography*, 3(2), pp. 117–144. doi:10.1093/jeg/3.2.117

Bathelt, H., Malmberg, A. & Maskell, P. (2004) Clusters and knowledge: Local buzz, global pipelines and the process of knowledge creation, *Progress in Human Geography*, 28(1), pp. 31–56. doi:10.1191/0309132504ph469oa

Becattini, G. (2002) From Marshall's to the Italian "industrial districts". A brief critical reconstruction, in: A. Q. Curzio & M. Fortis (Eds) *Complexity and Industrial Clusters: Dynamics and Models in Theory and Practice*, pp. 82–106 (Heidelberg: Physica-Verlag).

Berg, S. H. (2015) Creative cluster evolution: The case of the film and TV industries in Seoul, South Korea, *European Planning Studies*. doi:10.1080/09654313.2014.946645

Bergman, E. M. (2008) Cluster life-cycles: An emerging synthesis, in: C. Karlsson (Ed) *Handbook of Research on Cluster Theory*, pp. 114–132 (Cheltenham: Edward Elgar).

BMBF. (2009) Deutschlands Spitzencluster: Mehr Innovation. Mehr Wachstum. Mehr Beschäftigung. Bundesministerium für Bildung und Forschung, Referat "Neue Instrumente und Programme der Innovationsförderung"—Berlin.

Boggs, J. S. & Rantisi, N. M. (2003) The 'relational turn' in economic geography, *Journal of Economic Geography*, 3(2), pp. 109–116. doi:10.1093/jeg/3.2.109

Boschma, R. & Frenken, K. (2006) Why is economic geography not an evolutionary science? Towards an evolutionary economic geography, *Journal of Economic Geography*, 6(3), pp. 273–302. doi:10.1093/jeg/lbi022

Boyer, R. (1997) How does a new production system emerge?, in: R. Boyer & J.-P. Durand (Eds) *After Fordism*, pp. 1–55 (London: MacMillan).

Brenner, T. & Schlump, C. (2011) Policy measures and their effects in the different phases of the cluster life cycle, *Regional Studies*, 45(10), pp. 1363–1386. doi:10.1080/00343404.2010.529116

Dalum, B., Pedersen, C. O. R. & Villumsen, G. (2005) Technological life-cycles—Lessons from a cluster facing disruption, *European Urban and Regional Studies*, 12(3), pp. 229–246. doi:10.1177/0969776405056594

Dohse, D. (2007) Cluster-based technology policy: The German experience, *Industry and Innovation*, 14(1), pp. 69–94. doi:10.1080/13662710601130848

Duranton, G. & Puga, D. (2001) Nursery cities: Urban diversity, process innovation, and the life cycle of products, *American Economic Review*, 91(5), pp. 1454–1477. doi:10.1257/aer.91.5.1454

Eickelpasch, A. & Fritsch, M. (2005) Contests for cooperation—A new approach in German innovation policy, *Research Policy*, 34(8), pp. 1269–1282. doi:10.1016/j.respol.2005.02.009

Elola, A., Valdaliso, J., López, S. M. & Aranguren, M. J. (2012) Cluster life cycles, path dependency and regional economic development: Insights from a meta-study on Basque clusters, *European Planning Studies*, 20(2), pp. 257–279.

Enright, M. J. (2003) Regional clusters: What we know and what we should know, in: J. Bröcker, D. Dohse & R. Soltwedel (Eds) *Innovation Clusters and Interregional Competition*, pp. 99–129 (Berlin: Springer).

Feldman, M. P., Francis, J. & Bercovitz, J. (2005) Creating a cluster while building a firm: Entrepreneurs and the formation of industrial clusters, *Regional Studies*, 39(1), pp. 129–141. doi:10.1080/0034340052000320888

French, S., Leyshon, A. & Wainwright, T. (2011) Financializing space, spacing financialization, *Progress in Human Geography*, 35(6), pp. 798–819. doi:10.1177/0309132510396749

Fromhold-Eisebith, M. & Eisebith, G. (2008) Clusterförderung auf dem Prüfstand. Eine kritische Analyse, *Zeitschrift für Wirtschaftsgeographie*, 52(2–3), pp. 79–94.

Gancarczyk, M. (2015) Enterprise-and industry-level drivers of cluster evolution and their outcomes for clusters from developed and less-developed countries, *European Planning Studies*. doi:10.1080/09654313.2014.959811

Giuliani, E. (2013) Network dynamics in regional clusters: Evidence from Chile, *Research Policy*, 42(8), pp. 1406–1419. doi:10.1016/j.respol.2013.04.002

Grabher, G. (1993) The weakness of strong ties. The lock-in of regional development in the Ruhr area, in: G. Grabher (Ed) *The Embedded Firm*, pp. 255–277 (London: Routledge).

Granovetter, M. S. (1985) Economic-action and social-structure—The problem of embeddedness, *American Journal of Sociology*, 91(3), pp. 481–510. doi:10.1086/228311

Hassink, R. (2010) Locked in decline? On the role of regional lock-ins in old industrial areas, in: R. Boschma & R. Martin (Eds) *Handbook of Evolutionary Economic Geography*, pp. 450–468 (Cheltenham: Edward Elgar).

Hassink, R. & Shin, D. H. (2005) The restructuring of old industrial areas in Europe and Asia, *Environment and Planning A*, 37(4), pp. 571–580. doi:10.1068/a36273

Hervas-Oliver, J. L. & Albors-Garrigos, J. (2014) Are technology gatekeepers renewing clusters? Understanding gatekeepers and their dynamics across cluster life cycles, *Entrepreneurship and Regional Development*, 26(5–6), pp. 431–452. doi:10.1080/08985626.2014.933489

Ingstrup, M. B. & Damgaard, T. (2013) Cluster facilitation from a cluster life cycle perspective, *European Planning Studies*, 21(4), pp. 556–574.

Klepper, S. (2007) Disagreements, spinoffs, and the evolution of Detroit as the capital of the US automobile industry, *Management Science*, 53(4), pp. 616–631. doi:10.1287/mnsc.1060.0683

Lampel, J. & Meyer, A. D. (2008) Introduction: Field-configuring events as structuring mechanisms: How conferences, ceremonies, and trade shows constitute new technologies, industries, and markets, *Journal of Management Studies*, 45(6), pp. 1025–1035. doi:10.1111/j.1467-6486.2008.00787.x

Leamer, E. E. & Storper, M. (2001) The economic geography of the internet age, *Journal of International Business Studies*, 32(4), pp. 641–665. doi:10.1057/palgrave.jibs.84909988

Li, P. F., Bathelt, H. & Wang, J. C. (2012) Network dynamics and cluster evolution: Changing trajectories of the aluminium extrusion industry in Dali, China, *Journal of Economic Geography*, 12(1), pp. 127–155. doi:10.1093/jeg/lbr024

Link, A. N. & Scott, J. T. (2003) The growth of Research Triangle Park, *Small Business Economics*, 20(2), pp. 167–175. doi:10.1023/A:1022216116063

Livi, C., & Jeannerat, H. (2015) Born to be sold? Start-ups as products and new territorial lifecycles of industrialization, *European Planning Studies*. doi:10.1080/09654313.2014.960180

Longhi, C. (1999) Networks, collective learning and technology development in innovative high technology regions: The case of Sophia Antipolis, *Regional Studies*, 33(4), pp. 333–342. doi:10.1080/713693559

Lorenzen, M. (2005) Why do clusters change?, *European Urban and Regional Studies*, 12(3), pp. 203–208. doi: 10.1177/0969776405059046

Markusen, A. (1996) Sticky places in slippery space: A typology of industrial districts, *Economic Geography*, 72(3), pp. 293–313. doi:10.2307/144402

Martin, H., & Coenen, L. (2015) Institutional context and cluster emergence: The biogas industry in Southern Sweden, *European Planning Studies*. doi:10.1080/09654313.2014.960181

Martin, R. & Sunley, P. (2003) Deconstructing clusters: Chaotic concept or policy panacea?, *Journal of Economic Geography*, 3(1), pp. 5–35. doi:10.1093/jeg/3.1.5

Martin, R. & Sunley, P. (2006) Path dependence and regional economic evolution, *Journal of Economic Geography*, 6(4), pp. 395–437. doi:10.1093/jeg/lbl012

Martin, R. & Sunley, P. (2011) Conceptualizing cluster evolution: Beyond the life cycle model?, *Regional Studies*, 45(10), pp. 1299–1318. doi:10.1080/00343404.2011.622263

Maskell, P. (2001) Towards a knowledge-based theory of the geographical cluster, *Industrial and Corporate Change*, 10(4), pp. 921–943. doi:10.1093/icc/10.4.921

Maskell, P. & Malmberg, A. (2007) Myopia, knowledge development and cluster evolution, *Journal of Economic Geography*, 7(5), pp. 603–618. doi:10.1093/jeg/lbm020

Menzel, M.-P. & Fornahl, D. (2010) Cluster life cycles—Dimensions and rationales of cluster evolution, *Industrial and Corporate Change*, 19(1), pp. 205–238. doi:10.1093/icc/dtp036

Mossig, I. & Schieber, L. (2014) Driving forces of cluster evolution—Growth and lock-in of two German packaging machinery clusters, *European Urban and Regional Studies*. doi:10.1177/0969776414536061

Owen-Smith, J. & Powell, W. W. (2004) Knowledge networks as channels and conduits: The effects of spillovers in the Boston biotechnology community, *Organization Science*, 15(1), pp. 5–21. doi:10.1287/orsc.1030.0054

Piore, M. J. & Sabel, C. F. (1984) *The Second Industrial Divide: Possibilities for Prosperity* (New York: Basic Books).

Porter, M. E. (1998) Clusters and the new economics of competition, *Harvard Business Review* (November–December), pp. 77–90.

Potter, A. & Watts, H. D. (2014) Revisiting Marshall's agglomeration economies: Technological relatedness and the evolution of the Sheffield metals cluster, *Regional Studies*, 48(4), pp. 603–623. doi:10.1080/00343404.2012.667560

Saxenian, A. (1994) *Regional Advantage: Culture and Competition in Silicon Valley and Route 128* (Cambridge, MA: Harvard University Press).

Shin, D. H. & Hassink, R. (2011) Cluster life cycles: The case of the shipbuilding industry cluster in South Korea, *Regional Studies*, 45(10), pp. 1387–1402. doi:10.1080/00343404.2011.579594

Sinozic, T. & Tödtling, F. (2015) Adaptation and change in creative clusters: Findings from Vienna's new media sector, *European Planning Studies*. doi:10.1080/09654313.2014.946641

Skalholt, A. & Thune, T. (2014) Coping with economic crises—The role of clusters, *European Planning Studies*, 22(10), pp. 1993–2010.

Staber, U. & Sautter, B. (2011) Who are we, and do we need to change? Cluster identity and life cycle, *Regional Studies*, 45(10), pp. 1349–1361. doi:10.1080/00343404.2010.490208

Storper, M. (1985) Oligopoly and the product cycle—Essentialism in economic-geography, *Economic Geography*, 61(3), pp. 260–282. doi:10.2307/143561

Suire, R. & Vicente, J. (2009) Why do some places succeed when others decline? A social interaction model of cluster viability, *Journal of Economic Geography*, 9(3), pp. 381–404. doi:10.1093/jeg/lbn053

Ter Wal, A. L. J. & Boschma, R. A. (2011) Co-evolution of firms, industries and networks in space, *Regional Studies*, 45(7), pp. 919–933. doi:10.1080/00343400802662658

Tomlinson, P. R. & Branston, J. R. (2014) Turning the tide: Prospects for an industrial renaissance in the North Staffordshire ceramics industrial district, *Cambridge Journal of Regions Economy and Society*, 7(3), pp. 489–507. doi:10.1093/cjres/rsu016

Trippl, M., Grillitsch, M., Isaksen, A. & Sinozic, T. (2015) Perspectives on cluster evolution: Critical review and future research issues, *European Planning Studies*. doi:10.1080/09654313.2014.999450

Vernon, R. (1966) International investment and international trade in product cycle, *Quarterly Journal of Economics*, 80(2), pp. 190–207. doi:10.2307/1880689

Enterprise- and Industry-Level Drivers of Cluster Evolution and Their Outcomes for Clusters from Developed and Less-Developed Countries

MARTA GANCARCZYK

Faculty of Management and Social Communication, Institute of Economics and Management, Jagiellonian University in Krakow, Kraków, Poland

ABSTRACT *This article aims to discuss the international strategies of lead companies and the modularization of production networks as drivers of cluster evolution in developed countries, and to formulate propositions regarding the impact of those drivers on relationships with clusters in less-developed countries, based on literature reviews. Three streams of literature were combined, namely, that on (1) the role of lead companies in the development of industrial agglomerations, (2) the life cycle and evolution of clusters, founded on evolutionary economic geography and (3) the possibilities of upgrading by suppliers from less-developed countries. The article contributes by proposing a conceptual model that covers internal cluster evolution and the evolution of inter-cluster relationships globally to inform business and policy choices. Moreover, the research gap is addressed to study how the cluster dynamics in developed countries affect the upgrading opportunities for clusters in less-developed countries. The theoretical input consists in using the constructs of knowledge exploration and exploitation as mechanisms that determine cluster development prospects. Cluster development perspectives are shown as determined by those clusters' capacity to jointly pursue knowledge exploration and exploitation activities.*

1. Introduction

The explanation of drivers and outcomes of cluster change is necessary for planning regional innovation policy as well as strategies of regional enterprises. Recent studies in regional development and change indicate a refocus from territories and institutions as objects of analysis to investigating spatial dynamics from the perspective of an individual company growth (Frenken, 2007; Frenken & Boschma, 2007; Menzel & Fornahl, 2010,

Ter Wal & Boschma, 2011). Evolutionary economic geography holds that an individual company and its industry life cycle determine the life cycle of the entire regional agglomeration and focus on the internal forces leading to its evolution. These forces are described with the use of such theoretical constructs as co-evolution, path dependence, myopia and lock-in (Martin & Sunley, 2006; Maskell & Malmberg, 2007; Menzel & Fornahl, 2010; Ter Wal & Boschma, 2011). Another group of studies focuses on the external forces, specifically internationalization, cost and technology competition and digitalization of economic activity as determinants of cluster dynamics (Samarra, 2005; Alberti, 2006; Biggiero, 2006; Samarra & Belussi, 2006; Lorentzen, 2008; Propris et al., 2008). This perspective also puts emphasis on cluster lead companies as units of analysis, but here they are treated as agents affected mainly by these external forces instead of internal characteristics and processes taking place in the industrial agglomeration (Li et al., 2012). At the same time, in the current debate and empirical evidence, cluster is increasingly analysed as a part of global value chains with a focus on choices the cluster companies make when internationalizing, specifically about their linkages with other industrial agglomerations in less-developed, low-cost countries (Humphrey & Schmitz, 2004b; Gereffi et al., 2005; Saxenian, 2007). This translates into the opposition between upgrading and locking-in of clusters in developing countries, which attract foreign investment due to prevailing cost advantages. The latter phenomenon is affected by technological change in the form of modularization of production networks, which occurs at the level of industry.

Considering the above perspectives in the on-going discussion on cluster dynamics, in this article three streams of literature were combined, namely (1) growing literature on the role of lead companies in the development of industrial agglomerations, theoretically anchored in the international entrepreneurship and business, and strategic management concepts, (2) studies on the life cycle and evolution of clusters, founded on evolutionary economic geography and (3) literature on the possibilities of upgrading by suppliers from less-developed countries, based on the global value chain approach. These research perspectives call for integration into the theoretical framework that would cover internal cluster evolution and the evolution of inter-cluster relationships globally. We respond to this challenge by integrating two types of drivers of cluster evolution into one model of cluster dynamics. One type is industry-level technology change in the form of modularization of production networks that affects the spatial organization of production and innovation. This industry-level perspective provides relevant insights based on the premise that clusters undergo changes typical of their dominant industries (Abernathy & Utterback, 1978; Klepper, 1997). However, it should also be noted that development paths of clusters may differ from the general, industry path, due to the nexus of other factors, specific to each individual territorial system as path dependent (Martin & Sunley, 2006; Menzel & Fornahl, 2010). The cluster-specific factors are better reflected in another type of driver, which stems from individual strategies of lead companies and their networks. This approach differs from systemic analysis that treated clusters as groupings of homogenous firms, undergoing joint development paths together with the institutions of environment. Instead we put stress on entrepreneurial choices of specific options that are affected jointly, but not determined, by the cluster's internal characteristics (governance system and capabilities) and external forces (cost and technology competition and modularization). The micro-, individual-firm approach poses a threat of reductionism that does not capture the cluster's nature as a governance system implying synergies and untraded interdependencies. We argue that both the problems of pure micro- and pure industry-level analyses can be overcome by adopting the perspective of a company and its network. This approach will reflect the importance of individual entrepreneurial decisions and group mechanisms of path dependence and lock-in.

Latest contributions on the dynamics of industrial agglomerations concentrate prevailingly on clusters from developed countries.[1] The impact of those changes on clusters in less-developed countries, which absorb value chain activities relocated from the developed countries' clusters, requires deeper exploration, and this article attempts to identify this impact. Here, cluster evolution is discussed as a process affecting not only one entity but rather encompassing systemic changes among clusters worldwide and changes in the geography of innovation and production (Boschma & Fornahl, 2011).

The aim of the paper is to discuss international strategies of lead companies and modularization of production networks as drivers of cluster evolution in developed countries, and to formulate propositions regarding the impact of those drivers on relationships with clusters in less-developed countries, based on a literature review. In this article, cluster evolution is understood as a sequence of alterations in its structural characteristics that mark a new stage of development. It is assumed that cluster evolution can be analysed based on already formed industrial agglomerations, as nascent or emerging clusters are not clusters yet (Menzel & Fornahl, 2010). We recognize the importance of the knowledge of how clusters emerge; however, this contribution does not relate directly to clusters' life cycle, where the emergence is explicitly discussed as one of the phases.

The paper contributes by proposing a conceptual model of the internal cluster evolution and of the evolution of the relationships among clusters in countries with differing levels of development. The model encompasses drivers of cluster evolution in the form of modularization of production networks and of international strategies of the lead companies, which represent industry- and enterprise-level factors, accordingly. The theoretical background of the proposed model is built upon knowledge exploration and exploitation as mechanisms that determine cluster development prospects, and that are affected by network governance mode. By constructing this model we address the research gap of studying how the cluster dynamics in developed countries affect the upgrading opportunities for clusters in less-developed clusters. Moreover, the paper responds to the recommendation that network theory be included in the theory of cluster evolution (Ter Wal & Boschma, 2011). Here, we explicitly treat networks as a structural indicator of cluster change.

The discussion is structured into six sections, including the introduction and conclusion. In Section 2, the meaning of cluster evolution is proposed from the perspective of knowledge exploration and exploitation in the firms' networks. Sections 3 and 4 analyse the modularization of production networks as an industry-level driver, and international strategies of the lead companies as an enterprise-level driver of cluster evolution. Section 5 includes a proposal of the model of cluster evolution based on the earlier discussion. The conclusion follows.

2. Knowledge Exploration and Exploitation and Cluster Evolution

In this paper it is assumed that the cluster phenomenon and concept encompass concepts such as the specialized industrial agglomeration, industrial district or industrial production system (Vanhaverbeke, 2001; European Commission, 2002). We treat clusters as geographical concentrations of firms in one or a limited number of adjacent industries that form cooperative and competitive networks together with the institutions of environment (Porter, 1998; European Commission, 2002). Based on such an understanding, two structural properties that constitute the cluster phenomenon can be identified (Gancarczyk & Gancarczyk, 2013).

(1) Spatial and industrial concentration that implies a joint stock of knowledge and regional specialization (Piore & Sabel, 1984; Krugman, 1991; Porter, 1998; Bellandi, 2001).
(2) Network relationships among the companies and between companies and institutions of environment that form a governance system, conducive to knowledge and innovation development (Brusco, 1982; Pyke & Sengenberger, 1992; Putnam, 1995; Markusen, 1996; Saxenian, 2000; Asheim & Isaksen, 2003, pp. 36–40).

The dynamics of clusters can be analysed and assessed provided the meaning of its structural change is determined to mark new development stages. The structural change concerns shifts in the cluster structural characteristics, i.e. in its basic properties. Based on this position, shifts in the spatial and industrial concentration and in networking relationships would provide for a structural change (Gancarczyk & Gancarczyk, 2013) that delimits a new phase of cluster evolution. Such a change is of a long-term and durable nature, which differentiates it from a cyclical, temporal growth, or downturn in the regional and national economies (European Commission, 2002; Alberti, 2006; Biggiero, 2006; Zucchella, 2006). Structural change marks a specific developmental stage of an industrial agglomeration and thus identification of a sequence of those shifts for a specific cluster delineates its evolution.

Spatial and industrial concentration forms an objective basis for identifying clusters and it is normally analysed as shares of major industries and as a composition of industries related by input–output linkages. Geographical proximity and industrial concentration are a source of agglomeration externalities. Those benefits are passively achieved as not derived from the planned, purposeful organizational strategies but from spatial closeness of a critical mass of companies and organizations (Saxenian, 2000).

Networking activity represents an active exploitation of the benefits of spatial and industrial concentration by regional cluster participants. We assume a broad understanding of networks as a set of longer term cooperative and competitive relationships with selected partners (Johannisson, 1998). The importance of networks in generating innovations lies in their ability to transfer knowledge, not only information, as opposite to agglomeration effects (Maskell & Malmberg, 1999; Gertler, 2007). Internal networking is argued to be conducive for the success of industrial agglomerations in the majority of the research findings (Brusco, 1982; Pyke & Sengenberger, 1992; Putnam, 1995; Markusen, 1996; Saxenian, 2000; Asheim & Isaksen, 2003; pp. 36–40; Molina-Morales & Martínez-Fernández, 2006). On the other hand, linking the internal networks with international or global relationships is necessary to maintain competitiveness by new knowledge infusion and to avoid technological lock-in (Grabher, 1993; Glasmeier, 1994; Sorn-Friese & Sørensen, 2005).

Considering the nature of the core properties of clusters, it can be argued that changes in spatial and industrial concentration result from prior modifications in the networking structure, i.e. network evolution enables the explanation of why clusters change (Li *et al.*, 2012). Moreover, the cluster's competitive advantage is increasingly perceived as its ability to generate new knowledge and exploit the existing knowledge, which results in innovative activity. Therefore, clusters' evolution can be modelled based on the network perspective, with a special emphasis on how different forms of network impact knowledge exploitation and exploration to generate innovations. We argue that knowledge exploration and exploitation are mechanisms that explain clusters' innovative activities, and thus its evolution from growth to maturity and threat of decline, to renewal. Furthermore, they are discussed as theoretical constructs which are compatible with other concepts describing cluster evolution.

Exploration and exploitation of knowledge form a continuum of organization and network evolution. Knowledge exploration means discovering new knowledge, combining the pieces of existing knowledge into new combinations and creating new knowledge to provide innovations. Knowledge exploration is a source of product innovations, which result in new industries branching out of the core knowledge stock (March, 1991; Lee *et al.*, 2003; Danneels, 2012). Knowledge exploitation comprises transfer and utilization of the existing knowledge by expanding to new markets, applying economies of scale and scope and refining products and processes to achieve efficiency gains and incremental and process innovations. If not combined with knowledge exploration, such activities can lead to competency traps (Levinthal & March, 1993), rigid specialization and lock-in. Knowledge exploration involves more risk and higher level of investment than knowledge exploitation, however in the long term it is necessary to maintain a competitive advantage. The empirical evidence reports benefits from joint pursuit and balanced application of these two approaches (Hitt *et al.*, 2011; Sirén *et al.*, 2012). The possibility of exploiting the existing or exploring new knowledge depends largely on the governance of networking. Here, it is critical to discriminate between hierarchical and heterarchical relationships (Lam, 2007; Lorentzen, 2008). Exploring new knowledge mostly requires heterarchical relationships, in which power relations are uneven but authority and decisions decentralized, capabilities are dispersed and more balanced throughout network, reciprocity and interdependence govern both vertical and horizontal relations (de Propris *et al.*, 2008; Wall *et al.*, 2011). In consequence, hierarchical networks (centralized authority and capabilities, vertical relations, uneven power relations and dependence) generate mainly process innovations while heterarchical networks are conducive to product innovations (Lorentzen, 2008; de Propris *et al.*, 2008; Wall *et al.*, 2011). From the perspective of suppliers, network governance impacts their development prospects in terms of upgrading or lock-in (Gereffi, 1996; Humphrey & Schmitz, 2002, 2004a, 2004b; Gereffi *et al.*, 2005). Among the determinants of governance modes there are levels of technological complexity and formalization (codification) in the specific exchange, and supplier capabilities (Gereffi *et al.*, 2005). In buyer-driven networks (Gereffi, 1996), a lead company formulates requirements as to product characteristics and the supplier offers its own engineering and design to meet those expectations. Suppliers are directed at up-grading their processes and benefit from the advisory and technical support of the dominant companies (Winter, 2010). Similar relationships are present in modular networks (high technology complexity and codification, and high subcontractor capabilities) and relational networks (high technology complexity but low codification and high subcontractor capabilities) (Gereffi *et al.*, 2005). Captive or producer-driven networks denote hierarchical relationships between the dominant company and its suppliers that adopt its technology and follow specific standards (Gereffi, 1996; Humphrey & Schmitz, 2002; Gereffi *et al.*, 2005). It can be inferred that roles of suppliers in networks vary from those posing the threat of lock-in (hierarchical networks such as buyer-driven networks/ captive value chains) to those enabling the knowledge exchange and upgrading (heter-archical networks such as buyer-driven networks, relational value chains and modular value chains).

Theories of network and technology life cycles refer to different phenomena; however they demonstrate similar structural characteristics in terms of phases of development (Tushman & Rosenkopf, 1992; Glückler, 2007). Both of them are rooted in the organiz-ational theory of evolution (March, 1991; Levinthal & March, 1993), where the level of partner heterogeneity delimits the stages of the cycle and the prospects for innovative output. Specific stages of development for both technology and network overlap into the pattern from variation to selection to retention in different theoretical accounts (March, 1991; Tushman & Rosenkopf, 1992; Glückler, 2007). Variation is a stage in network development similar to knowledge exploration activity and it is normally charac-terized by heterogeneity of members and their industrial and technological profiles. Selec-tion and retention phases are marked by consistency of network membership, including their specialized industrial profiles (March, 1991; Glückler, 2007), which provides for knowledge exploitation activities. Radical and product innovations require networks of diverse partners and their technological heterogeneity, while incremental and process innovations can rely on less diverse partners, with more homogenous technological capa-bility (de Propris, 2002).

The dependence between knowledge and innovation development and network govern-ance mode is a major theme in the discussion on cluster dynamics, as reflected in the life cycle model and the evolution of spatial clustering in an industry.

The cluster life cycle model proposes the level of technological heterogeneity of firms as affecting the emergence, growth, sustainment, decline or renewal of industrial agglomerations (Menzel & Fornahl, 2010). The moderate technological focus of com-panies, accompanied by networks that are flexible and open to external knowledge, translates into product innovations and branching out to related industries, which revi-talize the agglomeration and provide for new growth prospects. Industrial agglomera-tions imply a regional specialization; however, in order to reflect the stages of development, it is imperative to differentiate between a rigid specialization, typical of the declining stage, and a moderate specialization, i.e. one that has a related diver-sification, observed in the growth and renewal (new growth) stages (Frenken *et al.*, 2007; Neffke *et al.*, 2011). Learning localized in the same region, however, outside of the cluster itself, is a source of the continuous technological variety and new growth. The positive dynamics depend either on path dependence, having its sources in the internal spin-off processes (Klepper, 2007), networks and local knowledge accumulations (Ter Wal & Boschma, 2011), or in more advanced development stages, are stimulated by external factors, including the behaviours of leading compa-nies (Maskell & Malmberg, 2007; Menzel & Fornahl, 2010). The complementary view on how to maintain variety and growth in the mature cluster is offered in the concept of the learning cluster (Hassink, 2005). The mechanism of revitalization and knowledge transfer consists in linking regional and international innovation systems, i.e. the knowledge sources need to be identified out of the cluster's region as well. It can be posited that heterogeneity of actors and open, flexible networks, stimulate exploration activities, which results in the related diversity and growth or renewal of the cluster. When technological profiles of actors become strongly focused and their networks closed, knowledge exploitation may lead to rigid specialization and the threat of decline. It can be inferred that growing and sustaining phases of the cluster life cycle are featured by a combination of knowledge exploration and exploitation activities.

The conception of industrial spatial clustering investigates the co-evolution of firms, industries and networks (Ter Wal & Boschma, 2011). Special emphasis is put on the networks which form a basis not only for knowledge spill-over process inside the cluster, but, due to their increasingly non-local nature, also for knowledge absorption and transfer out of the source cluster. Industry spatial clustering develops jointly with firms' capabilities and networking relationships. The framework follows the phases typical of the industrial agglomeration life cycle, however, it also takes into account the processes of relocation and either dispersal of the activities formerly concentrated in space or creation of the new clusters in other locations. Capabilities are considered as substantive, instrumental to knowledge exploitation and dynamic, higher order capabilities, which are necessary for knowledge exploration. Dynamic capabilities (Teece et al., 1997; Eisenhard & Martin, 2000; Zahra & George, 2002; Zollo & Winter, 2002; Di Stefano et al., 2010) mark the firm's competitive position in the changing environment, specifically, when the maturity phase follows and the threat of lock-in arises. Dynamic capabilities are demonstrated by absorptive capacity, the ability to change the position in the network, such as upgrading or locking-in, and they are critical to replicating the firm's routines when it relocates some or all of its activities to new geographical regions (Ter Wal & Boschma, 2011).

In the above discussion, the conceptual links between the firm's exploration and exploitation activities and cluster dynamics were identified. Table 1 presents how knowledge exploration and exploitation activities may translate into the specific phases of cluster development in the theoretical approaches analysed.

From the pattern-matching approach in Table 1, it can be inferred that clusters' positive dynamics depend on jointly pursuing the exploration and exploitation activities, which ensure new industries branching out and capitalizing on to-date products and services, accordingly. Exploration and exploitation underpin cluster dynamics, but on the other hand, they are dependent on the network characteristics. An optimal level of network technological diversity is recommended, i.e. not too low to prevent new industries branching out, and not too high to achieve economies of scale and scope in benefitting from already developed products. Moreover, open and flexible networks with more dispersed power and partner capabilities are preferable relative to rigid and highly centralized governance, which resonates with the earlier discussion on the importance of heterarchical networks for knowledge exploration and hierarchical networks for knowledge exploitation activities. In Sections 3 and 4, the evolution of cluster networks is analysed from the perspectives of the source clusters in developed countries and companies and clusters from less-developed countries. The possibilities for knowledge exploitation and exploration on the part of developed clusters as well as for up-grading or lock-in on the part of developing clusters will be pointed out, based on types of the networks established. Thus, the constructs of knowledge exploration and exploitation are adopted as mechanisms that determine cluster development prospects and that are dependent on the network governance mode.

Table 1. Companies' exploration and exploitation activities and cluster dynamics in (1) the cluster life cycle model and (2) the framework of the evolution of spatial clustering in an industry

Knowledge development	Cluster life cycle	Evolution of spatial clustering in an industry
Exploration	Emergence (technological heterogeneity of firms, low interaction)	Introductory stage (flexible, social networks, variety of firm capabilities)
Exploration combined with exploitation	Growth (focusing technology of firms, open and flexible networks)	Growth stage (stabilising core-periphery profile of networks, dense networks inside the cluster, possibility of stable and dense knowledge networks dispersed to other locations)
Exploitation with limited exploration	Sustainment (focused technology, open networks benefiting from synergies and external knowledge)	Maturity stage (stable core-periphery profile of networks, decreasing variety of firm capabilities due to shake-out)
Exploitation	Decline (strongly focused technology, closed networks impede cluster adaptability)	Industry decline (network rigidity, technological lock-in)
Exploration combined with exploitation	New growth (new technological heterogeneity, strong networks sourcing external knowledge)	The start of a new cycle (the importance of dynamic capabilities in relocating to new regions or in changing position in network)

3. Modularization as an Industry-Level Driver of Cluster Evolution

Cluster evolution is affected by life cycles of their dominant industries (Abernathy & Utterback, 1978; Klepper, 1997), however, it is not entirely determined by industry pathways, due to a nexus of other socio-economic and institutional factors affecting these territorial industrial systems (Martin & Sunley, 2006; Menzel & Fornahl, 2010). Therefore, the impact of industry patterns should be treated as a factor external to the cluster, which moderates the internal cluster dynamics.

Industry evolution is often perceived from the perspective of product and technology life cycles that translate into production systems and industry's spatial organization (Ter Wal & Boschma, 2011). The important technological factor associated with industry's development stages is modularization of production networks as observed in industries with complex, multi-element products, ranging from high technology (computers, consumer electronics and pharmaceuticals) to medium (auto components, machines, home appliance and plastic stationary) to low technology (toys and food) (Gangnes & van Assche, 2004; Lau, 2011). This pattern emerges after some level of codification of product parameters is achieved and the dominant design emerges, and it is normally associated with late-growth and maturing stages of industry development (Funk, 2009). It can be posited then, that the emergence of this pattern overlaps with the phase of agglomeration cycle when the industrial cluster had already been developed, i.e. it exhibits the features of a formed cluster, instead of an emerging one.

The modularization phenomenon can be analysed from at least four perspectives, namely as a (1) feature of complex systems, (2) technological pattern of product design, (3) system of production and (4) spatial organization of production and innovation.

Modularity is a feature of complex systems that facilitates change and evolution. The modular system includes subsystems (modules and functional elements) that adapt and change internally, independently of other modules, but are connected through codified interfaces to act as a whole. The intense interactions happen inside the modules, with limited interaction among modules, which helps to avoid side effects of improvements introduced in a specific subsystem (Frenken & Mendritzki, 2012).

Modularization as a technological pattern of product design consists in dividing the product into individual functional elements (modules and subsystems) that can be engineered separately, as their functional interdependencies are reduced (Sanchez & Mahoney, 1996; Baldwin & Clark, 2000; Lau, 2011). Instead, interfaces and functions of separate modules are specified and codified to enable the elements to act as a whole (Ulrich & Eppinger, 2000). These characteristics speed up the process of product improvements, as the innovation process is disintegrated into separate organizations responsible for specific modules. The modules are internally integrated and involve intense, tacit knowledge exchange to engineer and design, while codified and standardized information is exchanged between modules, with little interaction.

The modular pattern of product design stimulated the division of manufacturing into separate functional elements for which the individual companies could be responsible, and thus it caused the formation of modular production networks (Langlois, 2002; Sturgeon, 2002, 2003). As a system of production, modular production networks consist in the specialization of the companies in specific value chain activities of a complex product and in their horizontal integration within this specialization. The two dominant actors are lead companies responsible for product innovations, design, engineering and marketing and contract manufacturers (orchestrators, maestros and system integrators) responsible for process innovations, manufacturing and overall coordination of the value chain, including other suppliers. Here, we can see relationships of a heterarchical nature between the lead companies and contract manufacturers, and hierarchical links between contract manufacturers and suppliers of standardized components and services. In heterarchical relations, the lead companies benefit from focusing on high value activities, but are exposed to the threat of technology leakage and of losing connections with manufacturing, which often inspires product innovations (Sturgeon, 2002). Contract manufacturers assume a risky and capital-intensive manufacturing, but benefit from an opportunity to upgrade and vertically integrate to capture value from the lead companies. In the literature on global value chains, modular production networks are a latest type of production system (Sturgeon *et al.*, 2008), while the earlier developed types include vertical quasi-integration, relational networks of vertically integrated firms and relational, egalitarian networks in the form of flexible specialization (Humphrey & Schmitz, 2002; Sturgeon, 2002, 2003; Gereffi *et al.*, 2005). Referring to the earlier deliberations, systems that rely upon vertical or quasi-vertical integration are efficient in knowledge exploitation, while those governed by network modes of egalitarian, heterarchical nature are directed at knowledge exploration. Despite national origins, specific production systems are adapted in the integrated global economy and modified in different regional and national contexts. This forms a basis for convergence of production models on the global scale, but with some idiosyncratic alterations in the general, converged patterns on industrial, regional and national scales. According to this view, the latest wave of convergence of production systems is based on modular production networks (Gereffi *et al.*, 2005; Sturgeon *et al.*, 2008).

The two concurrent trends of (1) vertical disintegration and outsourcing by branded original equipment manufacturers and of (2) horizontal integration among companies dealing with manufacturing, process engineering and logistics, resulted in the new spatial organization of industry as well (Gangnes & van Assche, 2004; Baldwin, 2007; Lau, 2011). Namely, higher value-adding activities including R&D, product design, product development and high value/low volume or high-mix/medium- to high-volume manufacturing are concentrated in the clusters of developed countries (CDCs). These activities require the intense interaction and transfer of tacit knowledge. At the same time, due to cost pressures, lower value-adding activities, such as low value/low-mix/ high-volume manufacturing, are dispersed to lower cost locations or to clusters of less-developed countries (CLDCs) (Sturgeon, 2003; Gangnes & van Assche, 2004; Lee & Saxenian, 2008). The standardized technology is largely codified and thus, the information and knowledge flow do not require direct and intense interaction. As a consequence, agglomeration and concentration benefits prove less important, which facilitates relocation and dispersal of those activities out of the source CDCs. It is posited that upgrading opportunities for companies and clusters from developing countries are limited, due to limited access to the advanced knowledge (Ernst, 2004; Rugraff, 2010; Pavlínek, 2012). Based on this position, the global functional integration among clusters preserves the core-periphery organization of production and innovation. The process of relocating lower value-adding, manufacturing activities stimulates structural change in the source CDCs in terms of the level of industrial profile (higher value activities) and networking system within the cluster (heterarchical links with partners of high capabilities that deal with technological innovations). The links of the core clusters with CLDCs are prevailingly hierarchical and at arm's length, with little direct interaction due to high level of technology codification.

It can be argued, however, that the pattern: high value-added activities in developed countries and lower value activities in less-developed countries, should be modified considering the industry technology and capability base of network partners. In higher technology manufacturing and in knowledge intensive services such as information and communication technologies, finance and business, media and broadcasting (OECD, 1999; Mudambi, 2008; Huggins & Johnston, 2010), the pace of product innovations requires intensive search for talent and cost efficiencies worldwide. This leads to dispersal of higher value activities in terms of R&D to locations both in the developed and developing countries (Lam, 2007; Saxenian, 2007; Lee & Saxenian, 2008; Mudambi, 2008; Malecki, 2010; MacKinnon, 2012). The prevailing purpose is to decrease the cost of this activity, and more standardized elements in this area are relocated in the form of own subsidiaries or subcontracting to the less-developed economies. This activity can lead to up-grading and expanding the competence base in these economies and to the emergence of new competitors. The competitors assume similar locational strategies for R&D as their established counterparts. They set up facilities in the clusters with well-established reputation to pursue the process of catching up with the leaders, absorbing spill-overs through proximity and cooperation, and creating new industries through technology innovations (Mudambi, 2008). In the Taiwanese information technology industry, the firms passed the way from full product manufacturing to system integration as contract manufacturers in the 1980s, and further towards product design and even product development in the 1990s. This process was underpinned by intense collaboration and mobility between the technical community of multinational and Taiwanese firms, and increasingly non-hierarchical linkages between the companies. The upgrading in Taiwan clusters was, however, accompanied by relocation of lower value manufacturing to lower cost locations of mainland China (Lee & Saxenian, 2008).

The above analysis leads to *Proposition 1*, namely:

> The outcome of modularization of production networks is generally the concentration of higher value-adding activities in the developed countries' clusters and relocation of lower value activities to clusters of less-developed countries, however, the level of partner capability and industry technology have moderating effects. Upgrading of clusters from less-developed countries into higher value activities is stimulated by higher capability of these industrial agglomerations, supported by their heterarchical linkages with core clusters. This upgrading outcome is more plausible for clusters in advanced technology and knowledge intensive industries.

4. International Strategies of the Lead Companies as an Enterprise-Level Driver of Cluster Evolution

The important, firm-level factor of cluster evolution is internationalization and relocation of business activity undertaken by the lead companies in the established CDCs (Biggiero, 2006; Zucchella, 2006).This factor should be treated as internal to the cluster, relative to the industry-level driver of modularization discussed in Section 3. The logic of relocation processes from developed countries cannot be fully explained by modularization, as not all industries undergo modularization or are at a life cycle stage that enables adopting it. Moreover, beside general, industry-level technological and cost factors of relocation, there are also locational choices of an institutional and socio-cultural nature, which can be traced in the micro-perspective of an individual company and its network. Specifically, at this level of analysis we can observe how different types of network governance emerge to stimulate knowledge exploration and exploitation activities and thus the cluster development prospects.

Due to both cost and technology pressures from the general trend of internationalization, since the 1980s, the cluster lead companies passed from exporting to internationalization of the production process (Lorentzen, 2008). The phenomenon of the lead companies in clusters was observed not only in the agglomerations that assumed "hub-and-spoke" structures of small- and medium-sized companies centred around large companies (such as German or American agglomerations) but were also identified in the Italian industrial districts (Alberti *et al*., 2008; Munari *et al*., 2012). Several roles of leading companies as technology and international market gate-keepers or global pipe-lines were identified (Boschma & Ter Wal, 2007; Giuliani, 2011), even if those roles were often exercised by the companies of small to medium size (de Propris *et al*., 2008). The lead companies in CDCs assume classical cost or differentiation strategies, which involve changes in the industrial profile and relocation of the activities from the source cluster to foreign regions (Zucchella, 2006). Internationalization of the cluster lead companies can be treated as a cluster-specific driver of evolution, as choices and behaviours of those enterprises are at least to some extent dependent on their intra-regional networks and embedded relations (Biggiero, 2006).

In the cost strategy, selective relocation normally concerns lower value-adding activities such as assembling, manufacturing of standard components, distribution and processing. Higher value-adding activities such as R&D, marketing, advanced manufacturing, product and process design and engineering, and coordination of the entire value chain remain in the cluster. The options of selective relocation include outsourcing of raw materials and components, inward processing and foreign direct investments. Replicative relocation involves moving the majority of or the entire value chain of a good out of the source region (Biggiero, 2006; Zucchella, 2006; Semlinger, 2008). Both types of relocation mean losing business by to-date suppliers, which provides for considerable structural change in the level of spatial and industrial concentration, industrial profile and system of network relationships of the source clusters in developed countries. The specific options of cost strategy generate different types of networks and associated prospects for knowledge exploitation and exploration (Table 2), and, further, imply different development prospects for CDCs and CLDCs.

Selective relocation based on outsourcing raw materials and components, normally stimulates hierarchical networks with suppliers, where the dominant company establishes both technical parameters and terms of cooperation (Humphrey & Schmitz, 2004b; Gereffi *et al.*, 2005). Such relationships result in knowledge exploitation and threat of lock-in CDCs, with a limited possibility of upgrading by CLDCs. Inward processing is a more advanced form of cooperation. Despite these hierarchical relations, the higher competence of a supplier provides opportunities for changing the networks into more heterarchical ones and into those directed at knowledge exploration in the long run. The prospective outcomes include lock-in avoidance and sustainment of CDCs with possibilities for upgrading in CLDCs, as exemplified in the cooperation network between Veneto, Italy and Romanian firms in footwear and clothing industries (Crestanello & Tattara, 2011). In the case of selective relocation based on foreign investment (joint venture or green field), a corporate network is established that forms close links with new local environments. Such a strategic option enables heterarchical linkages that stimulate knowledge exploration and further growth of CDCs, with possible upgrading of their CLDC co-operators. Selective relocation is generally a positive strategy to maintain the competitive advantage by retaining the higher order competence within the cluster and by developing it through external

Table 2. The options of the cost strategy of internationalization of cluster lead companies in developed countries, types of international networks and prospects for knowledge exploration and exploitation

Option	Selective relocation			Replicative relocation
Networks and knowledge	Outsourcing of raw materials and components	Inward processing	Foreign direct investment (selected operations)	Foreign direct investment (the entire value chain)
Type of network	Hierarchical networks with suppliers	Hierarchical networks with subcontractors, possible heterarchical relations	Corporate hierarchical network, possible heterarchical relations	Corporate hierarchical network or hierarchy
Knowledge exploration/ exploitation	Exploitation	Exploitation, possible exploration	Exploitation and exploration	Exploitation

cooperation, which was observed in such cases as the Montebelluna clothing industry, Italy (Sammarra & Belussi, 2006) and Silicon Valley, USA (Saxenian, 2007).

Replicative relocation means losing connections with the source cluster and the knowledge stock stored there, while opening up to competition by imitation from rivals in lower cost locations. When adopted as a massive strategy it may bring decline of the source clusters such as in the case of silk confections near Como Lake (Alberti, 2006) and the clothing industry near Val Vibrata, Italy (Sammarra & Belussi, 2006). For lower cost locations, it brings the opportunity of either the emergence of new clusters or growth and upgrading of the existing ones. However, the sustainability of this positive dynamics is uncertain, as lead firms have only limited capability to appropriate the advantages of tacit knowledge and governance of their source clusters.

The differentiation strategies of internationalization are directed at achieving uniqueness through technological innovations and involve options such as focus on the global niches, development towards higher technology industries and complex products, and selective relocation (Biggiero, 2006; Sammarra & Belussi, 2006; Zucchella, 2006; Lam, 2007; Waxell & Malmberg, 2007; Gancarczyk & Gancarczyk, 2013). These options are intended for the retention of knowledge and the governance system of the source cluster, while restructuring into more advanced activities. They are less radical in changing the structural characteristics of CDCs in terms of the concentration level; however, they may impact the system of relationships and the industrial profile. Specific options discussed stimulate different types of networks and associated prospects for knowledge exploitation and exploration (Table 3) that bring development prospects for the core agglomerations and their CLDC counterparts.

Focus on the global niches for specialty products, such as branded Italian consumption goods, (Biggiero, 2006) is based on exporting and hierarchical networks with distributors, directed at knowledge exploitation. To a large extent, it preserves the existing knowledge and governance. However, in the long run it does not prevent the source clusters from the competence lock-in (Kalantaridis et al., 2011), as external collaboration in the production process or inputs is not directly assumed. As a consequence, upgrading of CLDCs cannot be considered here. Similar implications for network governance emerge from the other

Table 3. The differentiation options of internationalization strategy of cluster lead companies in developed countries, types of international networks and prospects for knowledge exploration and exploitation

Option			Selective relocation in the area of R&D	
Networks and knowledge	Focus on global niches	Development towards higher technology industries and complex products	Outsourcing of R&D projects	Direct investment in the area of R+R
Type of network	Hierarchical networks with distributors	Hierarchical networks with distributors	Heterarchical networks	
Knowledge exploration/ exploitation	Knowledge exploitation	Knowledge exploitation	Knowledge exploitation and exploration	

export-based option, namely development towards related industries with a higher technology level and product complexity. The examples include transition to production machinery and engineering in the ceramic tile industry in Emilia Romagna, Italy, transition to packaging machinery and materials in the food industry of Parma and Milano Italy, transition to software, environment technologies and biotechnology from computer hardware, electronics and precision instruments in Silicon Valley and Cambridge (Sammarra & Belussi, 2006; Zucchella, 2006) or from manufacturing to R&D and new sectors in small- and medium-sized enterprise-based Japanese clusters (Colovic, 2012). In order to prevent lock-in (Lam, 2007; Saxenian, 2007; Lorentzen, 2008), the two options discussed above may be combined with selective relocation. The latter consists in moving parts of the value chain, mainly those connected with R&D, to other locations by either outsourcing technology development projects or foreign direct investment (Lam, 2007; Waxell & Malmberg, 2007). It is directed at strategic coupling by purposeful selection of partners (MacKinnon, 2012) to form heterarchical relations that allow for knowledge exploitation and exploration. This sort of network governance opens up possibilities of new growth for the core clusters and for upgrading in agglomerations to which they relocate.

Relocation represents a challenge for companies that used to derive benefits from the locally embedded linkages and now need to establish new relationships. It is argued that there is a specificity of locational choices by CDC companies, which aim to achieve network and agglomeration benefits similar to those in the source cluster (Saxenian, 2007; Lorentzen, 2008; de Propris *et al.*, 2008). Namely, they search for partners in other clusters with related industrial profiles, which demonstrate cognitive, organizational, social and institutional proximity (Boschma, 2005), and their assets fit the lead companies' strategic needs (MacKinnon, 2012). The consequence is an emergence of the international networks of clusters with similar characteristics related by buyer–supplier links.

Companies from more competitive clusters divide work in the value chain with companies from developing clusters, which leads to the international specialization of clusters with similar industrial profiles (buyer–supplier relationships). As buyer clusters are normally assumed to have superior knowledge and market power relative to clusters of suppliers, the relationships will emerge as hierarchical, with a limited possibility for knowledge exchange. This conclusion should, however, be modified with regard to some options that involve technological collaboration with high capability suppliers, when heterarchical relations are formed. Besides supplier capability, the moderating factor in establishing the international network governance is industry (Kalantaridis *et al.*, 2011). This factor can be observed in the selective relocation of differentiation strategy, as typical of more technology advanced and knowledge intensive clusters, searching for collaboration with the counterparts of similar industrial profiles and capable of higher competence, both in the developed and less-developed clusters.

The above considerations lead us to *Proposition 2:*

The outcome of internationalization of cluster lead companies in developed countries is prevailingly a hierarchical division of work among clusters, however, the level of partner capability and industry technology have moderating effects. Industrial agglomerations in developed countries establish heterarchical linkages with clusters from less-developed countries that demonstrate a strong capability base, which results in upgrading of the latter clusters. This upgrading outcome is more plausible for clusters in advanced technology and knowledge intensive industries.

Proposition 2 complements Proposition 1 that referred to the effect of modularization on the division of higher value and lower value-adding activities between the clusters in countries with differing levels of development. Namely, it further describes this division as underpinned by hierarchical governance, and the discussion in the current section explains how this governance is established and what consequences it brings for cluster development prospects. Moreover, it refers to similar moderating factors in terms of the level of capability demonstrated by the partners from less-developed countries and technology level of the clusters considered. Overall, higher capability and advanced technology provide better prospects for establishing heterarchical linkages among clusters, which supports upgrading processes in CLDCs.

5. A Model of Enterprise- and Industry-Level Drivers of Cluster Evolution

Below is proposed a model that integrates industry-level (modularization) and enterprise-level (strategies of the lead companies) drivers of cluster evolution and their implications for development prospects of CDCs and upgrading of CLDCs and for relationships among these clusters (Figure 1).

An external factor of modularization of production networks forms conditions of industry spatial organization, as reflected in Proposition 1, which points to a concentration of higher value-adding activities in CDCs and a dispersal of lower value-adding activities to lower cost locations and CLDCs, with moderating effects of CLDCs' capabilities and industry technology. Namely, the higher the level of capabilities and technology advancement of industry, the better opportunities for upgrading of CLDCs, which may impact the dispersal of higher value-adding activities to clusters of less-developed economies, as well. The modularization, together with the international competition in the area of cost and technology, stimulates the behaviours of the lead companies from CDCs. They respond with specific strategies that differentiate by the choice of the industrial profile and type of relocation. From the CDCs' perspective, successful strategies depend on combining internal cluster networks with the external linkages. This translates into benefits of (1) preserving knowledge embedded in the networks of the source cluster and of (2) cost advantages and technology development through external sourcing. Positive dynamics of industrial agglomeration depend on jointly pursuing knowledge exploration and exploitation activities to avoid lock-in. This, in turn, is conditioned by the type of governance established with CLDC partners. From the point of view of clusters of developing economies, their upgrading and positive dynamics are conditioned by the capabilities they demonstrate, the strategies of CDC's leaders and the technological level of the industry in which they cooperate. Considering the benefits of both CDCs and CLDs, the most efficient long-term strategies are those based on selective relocation that enables heterarchical relations, namely selective relocation in differentiation strategy and selective relocation in cost strategy, which is based on inward processing and foreign direct investment. Selective relocation in the form of outsourcing and differentiation as focus or development towards more advanced activities, prevent positive effects due to their inward orientation or/and hierarchical governance. Replicative relocation in turn, hollows out the source agglomeration, while providing possibilities to upgrade by CLDCs. The upgrading and knowledge

outcomes

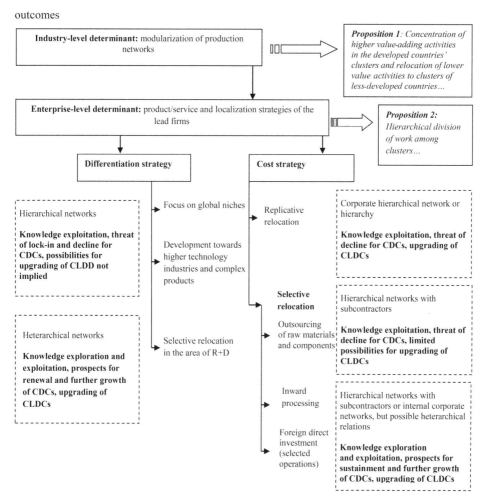

Figure 1. A model of enterprise and industry-level drivers of cluster evolution and their outcomes.

exploration of this strategy are impeded by losing the stock of knowledge that once belonged to the networks of the source cluster, which cannot be fully appropriated by the lead companies. The overall effect of the internationalization strategies of CDCs' leaders is reflected in Proposition 2, which states a hierarchical division of work among clusters which, however, is moderated by the CLDC partner's capabilities and industry technology, the latter factors stimulating more heterarchical links and upgrading by CLDCs.

6. Conclusion

In this paper three streams of literature were combined to synthesize on cluster dynamics, namely literature on (1) the role of lead companies in the development of industrial agglomerations, (2) the life cycle and evolution of clusters, founded on evolutionary economic geography and (3) the possibilities of upgrading by suppliers from less-developed countries. Based on these research streams, the industry-level and enterprise-level factors of cluster evolution were differentiated. The enterprise-level internationalization strategies were considered cluster-specific, while the industry-level factors were treated as affecting the cluster but external to it, with a more limited possibility to influence or control from the perspective of cluster participants or regional policy-makers. The concept of structural change was proposed to capture the considerable alterations that mark evolutionary stages within the cluster. The cluster evolution stimulated by industry- and enterprise-level factors was discussed from the perspective of network change. The research approach to synthesizing on cluster evolution was based on the logic that strategies of cluster lead companies create governance patterns which stimulate knowledge exploration and/or exploitation in the networks affected, from which cluster development prospects can be deduced. Clusters' development perspectives are thus shown as determined by their capacity to jointly pursue knowledge exploration and exploitation, which are determinants of innovative activity. The analysis resulted in the two propositions as to the outcomes of the drivers discussed on relationships among clusters worldwide. Following the propositions, the conceptual model that integrated internal (enterprise-level) and external (industry-level) drivers of cluster evolution and their outcomes for individual cluster change and for international relationships among clusters was proposed. The resultant conceptual framework opens up several practical and policy implications as to ways of stimulating exploration and exploitation activities of companies. Possible recommendations that stem from the proposed model include selecting specific strategies for international expansion as well as governance modes and production systems to implement those strategies. Further contributions that adopt this perspective may also refer to the achievements of organizational theory, strategic management and systems theory to study the interplay between company behaviour and governance modes in shaping structural change in clusters.

This article takes a broad view to look for some generalizations on cluster evolution. It is however, inconclusive and suggestive rather than providing definite responses to the complex issue of cluster dynamics. It also suffers from some limitations, such as relying on case studies with differing methodologies and focusing on a broad but not exhaustive range of industries to fully capture the influence of technology on various paths of cluster evolution. Moreover, the internal characteristics of clusters located in less-developed countries is a missing element and requires deeper investigation in future research. Despite assuming a cluster-specific perspective of lead companies, the paper still does not sufficiently discuss socio-economic and cultural issues that provide for divergent paths of cluster development. It should be asked, to what extent development paths of clusters are explained by those soft factors, instead of cost, technology and industry influences. For further research, it is also suggested that the model proposed in this article is operationalized into an appropriate methodology of empirical study.

Acknowledgments

I would like to thank Robert Hassink, Dirk Fornahl and Max-Peter Menzel for giving me an opportunity to contribute to this special issue. I am also grateful to two anonymous reviewers for their insightful comments.

Note

1. In this paper the notions of "developed" and "developing" or "less–developed" countries do not reflect any specific classification of country development as provided by the international institutions such as OECD or World Bank. The antinomy of "developed" versus "developing" or "less-developed country" denotes an uneven level of economy advancement between trading partners. In the empirical literature applied in this study clusters from "developing" countries encompass such examples as newly developed economies of Czechia, developing economy of Romania, emerging economies of India or China or developing/moderately developed economy of Poland.

References

Abernathy, W. J. & Utterback, J. M. (1978) Patterns of industrial innovation, *Technology Review*, 80(7), pp. 40–47.

Alberti, F. G. (2006) The decline of the industrial district of Como: Recession, relocation or reconversion?, *Entrepreneurship & Regional Development*, 18(6), pp. 473–501.

Alberti, F. G., Sciascia, S., Tripodi, C. & Visconti, F. (2008) *Entrepreneurial Growth in Industrial Districts* (Cheltenham: Edward Elgar).

Asheim, B. T. & Isaksen, A. (2003) SMEs and the regional dimension of innovation, in: B. T. Asheim, A. Isaksen, C. Nauvelaers & T. Tödtling (Eds) *Regional Innovation Policy for Small-Medium Enterprises*, pp. 49–77 (Cheltenham: Edward Elgar Publishing).

Baldwin, C. Y. (2007) Where do transactions come from? Modularity, transactions, and the boundaries of firms, *Industrial and Corporate Change*, 17(1), pp. 155–195.

Baldwin, C. Y. & Clark, K. B. (2000) *Design Rules* (Cambridge: MIT Press).

Bellandi, M. (2001) Local development and embedded large firms, *Entrepreneurship & Regional Development*, 13(3), pp. 189–210.

Biggiero, L. (2006) Industrial and knowledge relocation strategies under the challenges of globalization and digitalization: The move of small and medium enterprises among territorial systems, *Entrepreneurship & Regional Development*, 18(6), pp. 443–471.

Boschma, R. A. (2005) Proximity and innovation: A critical assessment, *Regional Studies*, 39(1), pp. 61–74.

Boschma, R. A. & Fornahl, D. (2011) Cluster evolution and a roadmap for future research, *Regional Studies*, 45(1), pp. 1295–1298.

Boschma, R. A. & Ter Wal, A. L. J. (2007) Knowledge networks and innovative performance in an industrial district: The case of a footwear district in the south of Italy, *Industry and Innovation*, 14(2), pp. 177–199.

Brusco, S. (1982) The Emilian model: Productive decentralisation and social integration, *Cambridge Journal of Economics*, 6(2), pp. 167–184.

Colovic, A. (2012) Territorial systems and relocation: Insights from eight cases in Japan, *Entrepreneurship and Regional Development*, 24(7–8), pp. 589–617.

Crestanello, P. & Tattara, G. (2011) Industrial clusters and the governance of the global value chain: The Romania-Veneto network in footwear and clothing, *Regional Studies*, 45(2), pp. 187–203.

Danneels, E. (2012) Second-order competences and Schumpeterian rents, *Strategic Entrepreneurship Journal*, 6(1), pp. 42–58.

Di Stefano, G., Peteraf, M. & Verona, G. (2010) Dynamic capabilities deconstructed. A bibliographic investigation into the origins, development and future directions of the research domain, *Industrial & Corporate Change*, 19(4), pp. 1187–1204.

Eisenhardt, K. M. & Martin, J. (2000) Dynamic capabilities: What are they?, *Strategic Management Journal*, 21(10–11), pp. 1105–1121.

Ernst, D. (2004) *How sustainable are benefits from global production networks? Malaysia's upgrading prospects in the electronics industry*. Working paper 57, East-West Center, Honolulu, June.

European Commission (2002) *Regional clusters in Europe, Observatory of European SMEs*, No. 2 (Luxembourg: Office for Official Publications of the European Communities).

Frenken, K. (ed.) (2007) *Applied Evolutionary Economics and Economic Geography* (Cheltenham: Edward Elgar).

Frenken, K. & Boschma, R. A. (2007) A theoretical framework for evolutionary economic geography: Industrial dynamics and urban growth as a branching process, *Journal of Economic Geography*, 7(5), pp. 635–649.

Frenken, K. & Mendritzki, S. (2012) Optimal modularity: A demonstration of the evolutionary advantage of modular architectures, *Journal of Evolutionary Economics*, 22(5), pp. 935–956.

Frenken, K., Van Oort, F. G. & Verburg T. (2007) Related variety, unrelated variety and regional economic growth, *Regional Studies*, 41(5), pp. 685–697.

Funk, J. L. (2009) The co-evolution of technology and methods of standard setting: The case of the mobile phone industry, *Journal of Evolutionary Economics*, 19(1), pp. 73–93.

Gancarczyk, M. & Gancarczyk, J. (2013) Structural change in industrial clusters—scenarios and policy implications, *Studia Regionalia*, 35, pp. 111–128.

Gangnes, B. & van Assche, A. (2004) *Modular production networks in electronics.* SMU Economics & Statistics working paper series, Singapore Management University, Singapore.

Gereffi, G. (1996) Global commodity chains: New forms of coordination and control among nations and firms in international industries, *Competition and Change*, 1(4), pp. 427–439.

Gereffi, G., Humphrey, J. & Sturgeon, T. (2005) The governance of global value chains, *Review of International Political Economy*, 12(1), pp. 78–104.

Gertler, M. S. (2007) Tacit knowledge in production networks: How important is geography?, in: K. R. Polenske (Ed) *The Economic Geography of Innovation*, pp. 87–111 (Cambridge: Cambridge University Press).

Giuliani, E. (2011) Role of technological gatekeepers in the growth of industrial clusters: Evidence from Chile, *Regional Studies*, 45(10), pp. 1329–1348.

Glasmeier, A. (1994) Flexible districts, flexible regions? The institutional and cultural limits to districts in an era of globalization and technological paradigm shift, in: A. Amin & N. Thrift (Eds) *Globalization, Institutions and Regional Development in Europe*, pp. 118–146 (Oxford: Oxford University Press).

Glückler, J. (2007) Economic geography and the evolution of networks, *Journal of Economic Geography*, 7(5), pp. 619–634.

Grabher, G. (1993) The weakness of strong ties: The lock-in of regional development in the Ruhr area, in: G. Grabher (Ed) *The Embedded Firm: On the Socioeconomics of Industrial Network*, pp. 124–165 (London: Routledge).

Hassink, R. (2005) How to unlock regional economies from path dependency? From learning region to learning cluster, *European Planning Studies*, 14(3), pp. 521–535.

Hitt, M. A., Ireland, R. D., Sirmon, D. G. & Trahms, C. (2011) Strategic entrepreneurship: creating value for individuals, organizations, and society, *Academy of Management Perspectives*, 25(2), pp. 57–75.

Huggins, R. & Johnston, A. (2010) Knowledge flow and inter-firm networks: The influence of network resources, spatial proximity and firm size, *Entrepreneurship & Regional Development*, 22(5), pp. 457–484.

Humphrey, J. & Schmitz, H. (2002) How does insertion in global value chains affect upgrading in industrial clusters?, *Regional Studies*, 36(9), pp. 1017–1027.

Humphrey, J. & Schmitz, H. (2004a) Governance in global value chains, in: H. Schmitz (Ed) *Local Enterprises in the Global Economy*, pp. 95–109 (Cheltenham: Edward Elgar).

Humphrey, J. & Schmitz, H. (2004b) Chain governance and upgrading: Taking stock, in: H. Schmitz (Ed) *Local Enterprises in the Global Economy*, pp. 349–381 (Cheltenham: Edward Elgar).

Johannisson, B. (1998) Personal networks in emerging knowledge-based firms: Spatial and functional patterns, *Entrepreneurship & Regional Development*, 10(4), pp. 297–312.

Kalantaridis, C., Vassilev, I. & Fallon, G. (2011) Enterprise strategies, governance structure and performance: A comparative study of global integration, *Regional Studies*, 45(2), pp. 153–166.

Klepper, S. (1997) Industry life cycles, *Industrial and Corporate Change*, 6(1), pp. 145–182.

Klepper, S. (2007) Disagreements, spinoffs, and the evolution of Detroit as the capital of the U.S. automobile industry, *Management Science*, 53(4), pp. 616–631.

Krugman, P. (1991) *Geography and Trade* (Cambridge, MA: Leuven University Press; Leuven: MIT Press).

Lam, A. (2007) Multinationals and transnational social space for learning: Knowledge creation and transfer through global R&D networks, in: K. R. Polenske (Ed) *The Economic Geography of Innovation*, pp. 157–189 (Cambridge: Cambridge University Press).

Langlois, R. (2002) Modularity in technology and organization, *Journal of Economic Behavior and Organization*, 49(1), pp. 19–37.

Lau, A. (2011) Critical success factors in managing product modular production design: Six company case studies in Hong Kong, China, and Singapore, *Journal of Engineering and Technology Management*, 28(3), pp. 168–183.

Lee, J., Lee, J. & Lee, H. (2003) Exploration and exploitation in the presence of network externalities, *Management Science*, 49(4), pp. 553–570.

Lee, C.-K. & Saxenian, A. (2008) Co-evolution and coordination: A systemic analysis of the Taiwanese information technology industry, *Journal of Economic Geography*, 8(2), pp. 157–180.

Levinthal, D. A. & March, J. G. (1993) The myopia of learning, *Strategic Management Journal*, 14(4), pp. 95–112.

Li, P.-F., Bathelt, H. & Wang, J. (2012) Network dynamics and cluster evolution: Changing trajectories of the aluminium extrusion industry in Dali, China, *Journal of Economic Geography*, 12(1), pp. 127–155.

Lorentzen, A. (2008) Knowledge networks in local and global space, *Entrepreneurship & Regional Development*, 20(6), pp. 533–545.

MacKinnon, D. (2012) Beyond strategic coupling: Reassessing the firm-region nexus in global production networks, *Journal of Economic Geography*, 12(1), pp. 227–245.

Malecki, E. J. (2010) Global knowledge and creativity: New challenges for firms and regions, *Regional Studies*, 44(8), pp. 1033–1052.

March, J. G. (1991) Exploration and exploitation in organizational learning, *Organization Science*, 2(1), pp. 71–87.

Markusen, A. (1996) Sticky places in slippery space: A typology of industrial districts, *Economic Geography*, 72(3), pp. 293–313.

Martin, R. & Sunley, P. (2006) Path dependence and regional economic evolution, *Journal of Economic Geography*, 6(4), pp. 395–437.

Maskell, P. & Malmberg, A. (1999) Localised learning and industrial competitiveness, *Cambridge Journal of Economics*, 23(2), pp. 167–185.

Maskell, P. & Malmberg, A. (2007) Myopia, knowledge development and cluster evolution, *Journal of Economic Geography*, 7(5), pp. 603–618.

Menzel, M.-P. & Fornahl, D. (2010) Cluster life cycles —dimensions and rationales of cluster evolution, *Industrial and Corporate Change*, 19(1), pp. 205–238.

Molina-Morales, F. X. & Martínez-Fernández, M. T. (2006) Industrial districts: Something more than a neighbourhood, *Entrepreneurship & Regional Development*, 18(6), pp. 503–524.

Mudambi, R. (2008) Location, control and innovation in knowledge intensive industries, *Journal of Economic Geography*, 8(5), pp. 699–725.

Munari, F., Sobrero, M. & Malipiero, A. (2012) Absorptive capacity and localized spillovers: Focal firms as technological gatekeepers in industrial districts, *Industrial and Corporate Change*, 21(2), pp. 429–462.

Neffke, F., Henning, M., Boschma, R., Lundquist, K.-J. & Olander, L.-O. (2011) The dynamics of agglomeration externalities along the life cycle of industries, *Regional Studies*, 45(1), pp. 49–65.

Organisation for Economic Co-operation and Development (OECD) (1999) *The Knowledge Based Economy: A set of Facts and Figures* (Paris: OECD).

Pavlínek, P. (2012) The internationalization of corporate R&D and the automotive industry R&D of east-central Europe, *Economic Geography*, 88(3), pp. 279–310.

Piore, M. J. & Sabel, C. (1984) *The Second Industrial Divide* (New York: Basic Books).

Porter, M. E. (1998) Clusters and the new economics of competition, *Harvard Business Review*, 76(6), pp. 77–90.

de Propris, L. (2002) Types of innovation and inter-firm co-operation, *Entrepreneurship and Regional Development*, 14(4), pp. 337–353.

de Propris, L., Menghinello, S. & Sugden, R. (2008) The internationalization of production systems: Embeddedness openness and governance, *Entrepreneurship & Regional Development*, 20(6), pp. 489–493.

Putnam, R. (1995) Demokracja w dzialaniu, *Znak*, Krakow.

Pyke, F. & Sengenberger, W. (1992) *Industrial Districts and Local Economic Regeneration* (Geneva: International Institute for Labour Affairs).

Rugraff, E. (2010) Foreign direct investment (FDI) and supplier-oriented upgrading in the Czech motor vehicle industry, *Regional Studies*, 44(5), pp. 627–638.

Sammarra, A. (2005) Relocation and the international fragmentation of industrial districts value chain: Matching local and global perspectives, in: F. Belussi & A. Samarra (Eds) *Industrial districts, relocation and the governance of the global value chain*, pp. 61–70 (Padua: CLEUP).

Sammarra, A. & Belussi, F. (2006) Evolution and relocation in fashion-led Italian districts: Evidence from two case-studies, *Entrepreneurship & Regional Development*, 18(6), pp. 543–562.

Sanchez, R. & Mahoney, J. T. (1996) Modularity, flexibility, and knowledge management in product and organization design, *Strategic Management Journal*, 17(Winter special issue), pp. 63–76.

Saxenian, A. (2000) Regional networks in Silicon Valley and Route 128, in: Z. J. Acs (Ed) *Regional Innovation, Knowledge, and Global Change*, pp. 123–138 (London: Pinter).

Saxenian, A. (2007) Brain circulation and regional innovation: The Silicon Valley-Hsinchu-Shanghai triangle, in: K. R. Polenske (Ed) *The Economic Geography of Innovation*, pp. 190–212 (Cambridge: Cambridge University Press).

Semlinger, K. (2008) Cooperation and competition in network governance: Regional networks in a globalised economy, *Entrepreneurship & Regional Development*, 20(6), pp. 547–560.

Sirén, Ch. A., Kohtamäki, M. & Kuckertz, A. (2012) Exploration and exploitation strategies, profit performance and the mediating role of strategic learning: Escaping the exploitation trap, *Strategic Entrepreneurship Journal*, 6(1), pp. 18–41.

Sornn-Friese, H. & Sørensen, J. S. (2005) Linkage lock-in and regional economic development: The case of Øresund medi-tech plastics industry, *Entrepreneurship & Regional Development*, 17(4), pp. 267–291.

Sturgeon, T. J. (2002) Modular production networks: A new American model of industrial organization, *Industrial and Corporate Change*, 11(3), pp. 451–496.

Sturgeon, T. J. (2003) What really goes on in Silicon Valley: Spatial clustering and dispersal in modular production networks, *Economic Geography*, 3(2), pp. 199–225.

Sturgeon, T, Biesebroeck, J. V. & Gereffi, G. (2008) Value chains, networks and clusters: Reframing the global automotive industry, *Journal of Economic Geography*, 8(3), pp. 297–321.

Teece, D. J., Pisano, G. & Shuen, A. (1997) Dynamic capabilities and strategic management, *Strategic Management Journal*, 18(7), pp. 509–533.

Ter Wal, A. L. J. & Boschma, R. (2011) Co-evolution of firms, industries and networks in space, *Regional Studies*, 45(7), pp. 919–933.

Tushman, M. L. & Rosenkopf, L. (1992), Organizational determinants of technological change: Toward a sociology of technological evolution, in: L. L. Cummings & B. M. Staw (Eds), *Research in Organizational Behavior*, 14, pp. 311–347 (Greenwich: JAI Press).

Ulrich, K. T. & Eppinger, S. D. (2000) *Product Design and Development*, 2nd ed. (Boston: Irwin/McGraw-Hill).

Vanhaverbeke, W. (2001) Realizing new regional core competencies: Establishing a customer-oriented SME network, *Entrepreneurship & Regional Development*, 13(2), pp. 97–116.

Wall, R. S. & van der Knaap, G. A. (2011) Sectoral differentiation and network structure within contemporary worldwide corporate networks, *Economic Geography*, 87(3), pp. 267–308.

Waxell, A. & Malmberg, A. (2007) What is global and what is local in knowledge-generating interaction? The case of the biotech cluster in Uppsala, Sweden, *Entrepreneurship & Regional Development*, 19(2), pp. 137–159.

Winter, J. (2010) Upgrading of TNC subsidiaries: The case of the Polish automotive industry, *International Journal of Automotive Technology and Management*, 10(2–3), pp. 145–160.

Zahra, S. & George, G. (2002) Absorptive capacity: A review, reconceptualisation, and extension, *Academy of Management Review*, 27(2), pp. 185–203.

Zollo, M. & Winter, S. G. (2002) Deliberate learning and the evolution of dynamic capabilities, *Organization Science*, 13(3), pp. 339–351.

Zucchella, A. (2006) Local cluster dynamics: Trajectories of mature industrial districts between decline and multiple embeddedness, *Journal of Institutional Economics*, 2(1), pp. 21–44.

Born to be Sold: Start-ups as Products and New Territorial Life Cycles of Industrialization

CHRISTIAN LIVI & HUGUES JEANNERAT

Research group in territorial economy (GRET), Institute of Sociology – University of Neuchâtel, Neuchâtel, Switzerland

ABSTRACT *Territorial innovation models and policy practices traditionally tend to associate the emergence, resurgence and growth of start-ups with the development of local industries, either as industrial pioneers or as innovative spinoffs embedded in a regional production system. This approach is in line with a "life cycle" pattern of innovation and of industrialization marked by sequential waves of growth and decline, by technological renewal and by sectorial transitions. In a knowledge and financial economy characterized by combinatorial knowledge dynamics, by even shorter project-based innovations and by global financial and production networks, this approach is called into question. Through the case of Swiss medical technologies (Medtech), this paper highlights how local medtech start-ups' evolution is shaped, from its early phase on, by the corporate venture strategies of multinational companies. While the economic potential of start-ups was traditionally perceived in a longer run, they seem to be more often "born to be sold" today. New research avenues and policy issues are finally derived from this particular case to address territorial innovation and competitiveness in the future.*

Introduction

In Schumpeterian ontology, entrepreneurs personalize economic change by their capacity to create and exploit new production or market opportunities. Beyond an individual action, entrepreneurship is also perceived as a collective innovation process embedded in territorialized institutions, actor relations and evolutionary pathways. Widely investigated since the 1980s, the Silicon Valley model has played a large part in carrying out and legitimizing this ontology through idealized visions of it. At the same time, it has become a reference of territorial competitiveness advocated by the current policy discourses and practices.

33

Nowadays, the familiar figures of William Hewlett, David Packard, Steve Jobs, Bill Gates, Larry Page or Sergey Brin typify the iconic image of entrepreneurs developing a new idea, prototyping a new product and starting a new business in their "garage" (Audia & Rider, 2005). Start-ups symbolize contemporary entrepreneurship at the cross-roads of science and industry, embedded in regional networks of firms, research and education bodies and capital ventures (Florida & Kenney, 1988). They are analysed at the core of nascent industries forging the resilience of a flexible regional production system exporting innovative and competitive products to distant markets (Saxenian, 1990, 1991).

The aforementioned entrepreneurial projects, the image of "garage" entrepreneurs and start-ups, reflect a "life cycle" pattern of innovation and of industrialization marked by sequential waves of growth and decline, by technological renewal and by sectoral transitions. Often associated with the success stories of Hewlett-Packard, Apple, Microsoft or Google, start-ups are commonly viewed as the potential inception of a new industrial trajectory which is expected to grow, create jobs and disseminate innovation locally.

In a knowledge and financial economy characterized by combinatorial knowledge dynamics (Crevoisier & Jeannerat, 2009), by project-based innovations (Grabher, 2002a) and by global financial and production networks (Coe *et al.*, 2014), this idealized approach of regional innovation and growth is called into question. Through outsourcing and corporate venture strategies, multinational companies step in at an early stage of the enterprises' incubation (Chesbrough, 2002; Garel & Jumel, 2005; Ben Hadj Youssef, 2006). While local conditions of knowledge and capital transfers can be influential in the emergence of entrepreneurship (Kenney & Von Burg, 1999; Delgado *et al.*, 2010), the growth of a start-up seems to be more than ever bound to the decisions of global stakeholders. What kinds of entrepreneurship, ecosystems and evolutionary paths are implied by such interdependencies? How does it impact on theoretical and policy models of regional innovation and territorial competitiveness?

Through the case study of Swiss medical technologies (medtech), this paper sheds light on three different aspects of this question. Firstly, it is observed that medtech start-ups' trajectories are shaped by the two contrasting territorial dynamics of knowledge and financial anchoring. While fundamental technology and incubation capital build on local resources, industrial production and market exploitation take place, from its early phase onwards, through the investments of large listed multinational companies. Secondly, the entrepreneurial plan behind the creation of a new start-up indicates a fundamental change is underway. While the economic potential of start-ups was traditionally viewed as a longer term prospect, they are now "born to be sold". Corporate venture has become a strategic tool for large companies to tap external innovation processes taking place in the local milieu. Thirdly, the identification and evaluation by investors of potentially lucrative start-ups involve complex intermediation processes. From this point of view, international fairs and opinion leaders are the key means of justifying and legitimating the value of local medtech start-ups, purchased as a product.

Regional Innovation: Localized Entrepreneurship and Industrial Growth

In a post-Fordist era, the term innovation is commonly used to explain the success of particular enterprises, industries and regions facing production cost competition in the globalized economy. Inspired by Schumpeter's approach to economic change, many contemporary theories and policy discourses view innovation as a dialectical interplay of "emergence" and "growth" (Cooke *et al.*, 2011b).

On the one hand, economic change and evolution emerge from entrepreneurship (Rocha, 2004). Entrepreneurship is considered the fundamental socio-economic driver through which contingent resources (Bathelt & Glückler, 2005; Stam, 2010; Julien & Marchesnay, 2011) are turned into new products or production processes through creative destruction, production and recombination (Schumpeter, 1935). Emergence occurs through pioneer entrepreneurs or large incumbent firms breaking away from an existing market offering (e.g. a new product or a new use of an existing product) and/or an established production system (e.g. a new technology or a new supply chain).

On the other hand, growth in economic change is usually considered through the pattern of industrialization (Klepper, 1997; Schmitz, 1999; Chataway & Wield, 2000). For Schumpeter (1939, p. 98), innovations are not "isolated events": they "tend to cluster, to come about in bunches". We here consider industrialization, in a broad and fundamental definition, as the process by which related entrepreneurial projects and production are developed on an extensive scale. Industrialization does not restrain to intensive manufacturing or economies of scale. It more broadly characterizes the agglomerated growth (Hilhorst, 1998) achieved through "collective efficiency" and increasing returns derived from external economies and joint action in particular production systems (Schmitz, 1999). It may be driven by processes of dissemination (e.g. through knowledge spillovers or competition–cooperation dynamics), concentration (e.g. dedicated competences, workforces and infrastructures) or specialization (e.g. a specialized supply chain). Through productive, corporate or market growth, industrial development generates new employment and new commercial revenues in relation to product and process innovations (Klepper, 1997).

In regional studies, entrepreneurship is usually regarded as the capacity of local actors to foresee and undertake individual and collective projects in a changing environment, based on specific regional resources (e.g. social, cultural or technical capital) (Saxenian, 1994; Maillat, 1995; Thierstein & Wilhelm, 2001; Stam, 2007). Innovation develops endogenously within local production systems competing beyond regional boundaries (Coffey & Polèse, 1984). More operational approaches have subsequently viewed entrepreneurship as the ability of specific regional innovation systems to turn locally generated knowledge into successful entrepreneurial projects (Cooke, 2001; Doloreux, 2002) where entrepreneurs are considered not only as an individual pursuing a personal vision, but also as a social agent situated in a wider system of production (Scott, 2006, p. 4). Within this system, local venture capital investors are the key players providing capital resources and managing expertise and strategic directions in the development of nascent firms (Florida & Kenney, 1988; Kenney & Von Burg, 1999; Feldman, 2001).

In the past decades, the spatial dynamics of entrepreneurship and emerging innovations have been the subject of various analyses particularly focused on regional economic growth and clusters (Kenney & Von Burg, 1999; Feldman *et al.*, 2005; Kiese & Schätzl, 2008; Glaeser & Kerr, 2009; Delgado *et al.*, 2010; Trettin & Welter, 2011). In this context, clusters and entrepreneurship have become very popular subjects in regional science and economic geography (Sternberg & Litzenberger, 2004). According to Delgado *et al.* (2010, p. 500) and Sternberg and Litzenberger (2004, p. 770), clusters facilitate new business formation and the growth of successful start-ups by lowering the cost of entry, enhancing opportunities for innovation-based entry, allowing start-ups to leverage local resources to expand new businesses more rapidly and offering a positive regional environment. Reducing barriers to entry and growth and enhancing regional comparative advantage, the presence of a strong cluster environment should be a central factor of entrepreneurial vitality (Delgado *et al.*, 2010, p. 498). Not merely the result of individual efforts, entrepreneurship has been depicted as a collective process embedded in particular relational, institutional and evolutionary configurations situated in time and space (Feldman, 2001; Lambooy, 2005; Ferrary, 2008). Stam (2007) emphasizes that new enterprises creation is characterized by different evolutionary phases beginning with recognition of a new business opportunity by the entrepreneurs and ending in a "growth syndrome" represented by a decrease of a firm.

Not confined to the question of entrepreneurship, industrialization has also been addressed as a *sine qua non* condition of development within territorial innovation (Scott, 1986; Scott & Storper, 1992; Hilhorst, 1998; Schmitz & Nadvi, 1999). Not only have regional innovation systems been depicted as spatial contexts of emerging innovation, but they are also particular socio-economic arenas enabling entrepreneurial projects to "take off" and, at the same time, "anchoring" them through local clusters of activities (Porter, 1998; Cooke & Martin, 2006), enhancing knowledge dissemination, flexible specialization (Scott, 1988; Simmie, 2005), related innovations (Frenken & Boschma, 2007) or spatial agglomeration (Stam, 2007; Vatne, 2011). If regional clustering of innovation does not necessarily follow the trajectory of a single industry, its endogenous development bears an industrial dimension following the sequential homogenization phases of activities in particular industrial or thematic fields (Menzel & Fornahl, 2010).

This industrial aspect of clustering has given rise to debated models of territorial competitiveness, highlighting the regional advantage created by a "Marshalian" specialization (Moulaert & Sekia, 2003) or by a "Jacobian" diversification of local innovative activities (Cooke, 2008). Nevertheless, all these models share a common view: understanding territorial competitiveness is not only about pointing out how innovation emerges in a particular spatial context, but also about addressing how innovation generates new employment and revenue through export-based (basic) and induced (non-basic) activities in this territory (Polèse & Shearmur, 2009).

The Life Cycle Approach: A General Conception of Regional Development

The interplay between entrepreneurship and industrialization in regional innovation processes has traditionally been interpreted and conceptualized in a life cycle approach. Initially adopted by Marshall (1890), who compared the evolution of businesses in the nineteenth century with the birth, growth, maturity and death of trees in a forest, the "life cycle" metaphor has gained common currency in describing the organic nature of economic processes (Penrose, 1952). Used to describe the way in which firms and industries develop within the ecological context of technology, product and market selection, several approaches based on life cycle posit innovation as sequential waves of emergence, growth, maturity and decline (Vernon, 1966; Markusen, 1985; Klepper, 1997). Drawing upon a similar metaphor, numerous works have provided various interpretations of territorial competitiveness along with the different stages of development that particular regional production systems may face (Vernon, 1966; Stam, 2007; Menzel & Fornahl, 2010; Potter & Watts, 2011; Tichy, 2011; Cooke *et al.*, 2011a; Cooke *et al.*, 2011b).

In phases of emergence and growth, geographical proximity can provide relational (e.g. informal, multifunctional or specialized networking among regional economic and non-economic actors) and institutional facilities (e.g. routines and policy support) to stimulate new entrepreneurial projects and to overcome market or technological uncertainties related to the creation and industrialization of new market offerings (Stam, 2007; Carrincazeaux & Coris, 2011; Potter & Watts, 2011). This phase is characterized by two dominant processes related to the exploitation of new market opportunities and the delivery of products to a growing product market (Stam, 2007, p. 30). In the maturity phase, standardized technologies, production processes and markets become less dependent on a particular innovation milieu. Relocation of activities is easier and creates a new spatial division of labour in a global market (Vernon, 1966; Tichy, 2011). Increasingly based on extra-regional relations or global pipelines (Bathelt *et al.*, 2004; Isaksen, 2011), the stages of maturity and decline are usually not directly related to innovation-driven territorial competitiveness. Innovation may develop incrementally alongside a particular market positioning and sectoral trajectory, but competitiveness is primarily achieved through conservative principles (market oligopolies, technical and structural path dependencies) underlying a potential decline in the original production system through latent lock-in (Grabher, 1993; Boschma & Lambooy, 1999).

From this point of view, territorial innovation models have hitherto primarily focused on regional emergence and growth of innovation (Asheim & Coenen, 2005; Cooke *et al.*, 2011a). Phases of maturity and decline are usually regarded as inevitable aspects of new potential emergence through innovative diversification, adaptation or reconversion. Regional innovation systems do not necessarily draw upon a single product or sector. They usually build upon subsequent emergences and related industrial life cycle types (Cooke *et al.*, 2011b). In other words, regional innovation systems are fundamentally depicted as specific territorial contexts of entrepreneurial (re)emergence(s) and industrial growth through a local innovative valuation of resources (Figure 1).

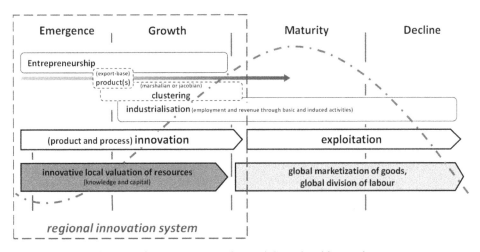

Figure 1. Innovative emergence and growth in regional innovation systems.
Source: Author's own work.

Innovation Policies and Contemporary Reflections

In the last decade, a plethora of public initiatives has been launched to enhance regional innovation and territorial competiveness. Usually taking the case of the Silicon Valley as a reference, "technopole" and "cluster" strategies are considered as policy best practices (Martin & Sunley, 2003; Tödtling & Trippl, 2005; Brenner & Schlump, 2011). In various aspects, these policies are implicitly rooted in a life cycle approach of regional entrepreneurship and growth.

On the one hand, these policies seek to stimulate pioneer entrepreneurs and incumbent companies by providing pre-competitive funding to R&D projects and start-up ventures. On the other hand, they aim to foster creative knowledge sharing and dissemination among regional actors in related fields of activities through proactive networking. Generally speaking, public intervention tends to be viewed ideally as the third component in a triple-helix scenario, whereby it provides "assistance" for the emergence and growth of "linear" innovations taking place between science and industry (Etzkowitz & Leydesdorff, 2000; Thierstein & Wilhelm, 2001; Etzkowitz, 2006).

Such initiatives tend to share an implicit view of innovation: successful start-ups are the embodiment of innovation (Feldman, 2001) facilitating sometimes nascent industries that will underlie the resilience and growth of a regional production system; supporting the emergence of local innovation today prepares the ground for the industries of tomorrow; new regional employments and revenues will come along with innovations. This stylized approach to regional innovation and growth policies can nevertheless be challenged. Drawing upon various seminal critiques addressed by recent debates in regional studies, three prominent and fundamental issues emerge.

Firstly, in a complex knowledge-based society, economic development and competitiveness are strongly driven by combinatorial knowledge dynamics (Gibbons *et al.*, 1994; Crevoisier & Jeannerat, 2009). Innovation increasingly tends to emerge across different sectoral life cycles rather than within single trajectories. Accordingly, regional innovation tends to emerge and develop through related varieties taking place across different local clusters and life cycles (Asheim *et al.*, 2011).

Secondly, in this context of knowledge-intensive innovation, entrepreneurship builds on permanent and shorter-run projects (Grabher, 2002b). Regions are, in this context, complex "project arenas" (Qvortrup, 2006) or "adaptive systems" (Martin & Sunley, 2011) of continuous innovative (re)emergence that have to overcome the path-dependent lock-ins inherent in long industrial waves and stable phases of industrial maturity. Furthermore, regional revenue is generated from knowledge-intensive activities, selling tailor-made solutions rather than export-based products.

Finally, territorial innovation processes are embedded in increasingly global production and financialized networks (Corpataux *et al.*, 2009; Coe *et al.*, 2014). In the traditional life cycle approach to territorial innovation, spatial division of labour was traditionally described as a "push" movement (Tichy, 2011): the relocation of activities and foreign direct investments are undertaken by mature companies from their home region towards specialized and lower-cost supplying regions. Nowadays, this traditional process is challenged by two fundamental phenomena. On the one hand, global economic financialization has increased the liquidity/mobility of capital, which can be instantly invested in distant and attractive listed businesses (Corpataux *et al.*, 2009). On the other hand, large multinational companies have become global investors in outsourced innovations through corporate venturing (Chesbrough, 2002; Garel & Jumel, 2005; Ben Hadj Youssef, 2006). In this context, relocations and foreign direct investments tend to occur at an earlier development phase in a "pull" process, which consists of "picking up" the competitive winners within global innovation networks.

How do these new challenges actually affect established models of regional innovation and territorial competitiveness? How should conventional policy and life cycle approaches be reconsidered in such a context? The next section examines these questions through the particular case of medical technologies in Western Switzerland and tries to give new keys to understanding the territorial and industrial dynamics of firm emergence and development.

The Medtech Industry in Western Switzerland

The medical innovation literature shows that new medical devices and applications arise out of interactions between different actors such as universities, hospitals, laboratories and enterprises (Gelijns & Thier, 2002) and emphasizes that structures shaping innovations can be distinguished from pharmaceuticals and biotechnology (Weigel, 2011, p. 45).

Considered to be the area of life sciences covering the various economic activities of research, subcontracting, development and marketing of medical devices and applications (Medtech Switzerland, 2012), medtech is one of the Swiss economy's flagship industries, generating around $5 billion per annum (Klöpper & Haisch, 2008, p. 11) and employing some 50,000 people (i.e. 1.1% of the country's workforce) (Medtech Switzerland, 2012, p. 31). This sector shows a sustained annual growth of 5–20% depending on the industrial branches (Fritschi, 2006).

The Swiss medical industry currently accounts for a total of 1600 companies[1] subdivided into manufacturers, producer suppliers, distributors and companies specializing in the supply of services to medical device producers (Medtech Switzerland, 2012, pp. 30–31). According to Klöpper and Haisch (2008), changes in the Swiss medical industry have primarily been driven by three factors. Firstly, medtech companies have benefitted from the Jura region's rich technological and research environment, thanks to the existing watchmaking industry there, producing various high-quality, high-precision components. Secondly, the high prices on the Swiss domestic market have often meant that there is money available for medical investment and innovation, helping local businesses to be more innovative than their market competitors. Thirdly, research by public laboratories, both basic and applied, has enabled the development of major medical projects within the country.

Territorially speaking, most businesses involved in medical work are based in the Zürich and Western Switzerland regions (Figure 2). With a long tradition in this field, the latter region has a dense population of medical actors (e.g. state-run hospitals and private clinics), research institutes (e.g. the university institute and private research investment), industry's major multinational companies (Klöpper & Haisch, 2008, p. 12) and a growing number of medtech start-ups (Figure 3).

Compared to the rich literature on innovations and clustering evolution in biotechnology, the existing literature on medical device industry is very limited (Weigel, 2011). This lack of literature raises some important issues for terminology and definition of "medtech". While the term "medtech" is commonly used by both public authorities to define a promising industry and entrepreneurs to designate their business activity, this

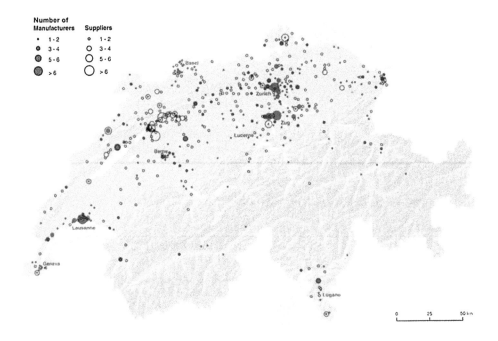

Figure 2. Number and localization of Medtech manufacturers and suppliers in Switzerland.
Source: Medtech Switzerland, 2012, p. 31.

term is inherently ambiguous. It can effectively be seen as both a service industry and a goods industry, as it represents not only those industries which produce medical devices but also those actors supplying services which are not identifiable with a specific product or technology. Henceforth, despite being implied by the term, "medtech" cannot be defined as a technology or by a clearly identifiable product type. For this reason, numerous actors in this field prefer to view medtech activities not as a specific technology or sector, but rather as a "market".

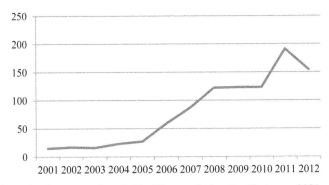

Figure 3. Medtech start-ups founded in Western Switzerland between 2001 and 2012.
Source: www.startup.ch, as per 12.12.2013.

These considerations raise several questions: how do we define current medtech activities? Would it be appropriate to speak of the emergence of a new medical cluster in Western Switzerland? How should we interpret and understand the emergence and development of these activities through the medtech start-ups that have recently sprung up in this region? What kind of critical reflections might the case of Western Switzerland's medtech industries bring to bear on the traditional cluster life cycle approach?

Method of Inquiry

These questions were examined in a case-study research (Merriam, 1998; Stake, 2005; Gerring, 2007; Yin, 2009) carried out between October 2012 and March 2013 as part of a project financed by the Fond National Suisse de Recherche Scientifique.[2] According to Merriam (1998), the case-study method provides insides into social phenomena in order to demonstrate their complexity and the context within which they were drafted. Therefore, it emerges from a process of interaction between information gathering, interpretations, literature and reporting (Yin, 2009). Searching to analyse the industrial evolution, socio-economic dynamics and territorial impacts of the medtech sector in Western Switzerland, our qualitative study was principally based on a mix method mobilizing semi-structured interviews, participant observation, panel of experts and an in-depth document analysis.

Five categories of actors were particularly identified (Table 1). The first type was the start-ups (a total of 16) developing new medical devices. According to Lebret (2007, p. 24), we considered "start-ups" as nascent firms, born from an entrepreneur's idea linked, or not, to an institutional actor (e.g. a university or a firm incubator) and having the possibility to become a larger enterprise. The second and the third types of actors were multinational enterprises (MNEs) and medium-sized enterprises (a total of 15) producing and selling medical devices in international markets. Fourthly, particular investors (a total of 5) were also interviewed to address the financial issues and rationales underlying the creation and the development of new firms. Finally, the fifth category is represented by professional associations and exhibition organizers, which provide a network creation between medical actors.

A total of 30 semi-directive interviews lasting one to two hours have been conducted in the Western Switzerland region with entrepreneurs, and representatives of small and medium enterprises (SMEs), of MNEs, of investors and of professional associations. The interviewees were selected according to a theoretical sampling (Guillemette & Luckerhoff, 2009), seeking to explore the general characteristics of medtech entrepreneurs, their networks, their enterprises and their regional involvement. The particular history of the firms, of their products and of their funding was examined. An in-depth web and document analysis was also realized to gather information about innovative projects, initiatives and experts' views related to medtech activities in Western Switzerland and abroad.

Once a saturation of results was achieved,[3] all interviews were transcribed and subjected to qualitative data analysis[4] (Corbin & Strauss, 2008; Silverman, 2010; Grbich, 2012). Our qualitative data were analysed through a conceptual coding (Nagy Hesse-Biber & Leavy, 2011). A descriptive coding of the textual and contents data generated by interviews and documents allowed us to build analytical categories to interpret three main issues: (1) the main changes in the medtech sector over the last years in Western Switzerland, (2) the

Table 1. Field actors and their principal functions[a]

Categories of actors	Semi-structured interviews and document analysis	Document analysis and participant observations	Function of actors	Examples
Start-ups	Aleva Neutheraputics, Sensimed, Melebi, Medos, Odus Technologies, Perfusal and other 5 medtech start-ups	5 local medtech start-ups	Develop a new application trying to solve some medical problems	Development of a new soft medical device able to treat diabetes diseases
MNEs	Medtronic, Phonak and other 3 MNEs	1 MNEs	Develop and sell products at a global scale	Worldwide production and sale of medical devices or application (e.g. hearing aid)
SMEs	Symbios, Valtronic, BienAir, Oscimed and other 3 SMEs	2 local SMEs	Develop, produce and sell product at a more regional/ multinational scale	Development and production of a new sort of prosthesis to sell in the region and in some other occidental countries
Investors	Capital proximité, and other 3 investors	1 investors	Support start-ups creation and development in order to have a capital gain	Monetary support during an early stage of new medtech firms
Associations and expositions	Medi SIAMS, and other 2 associations/ expositions	NEODE, Forum Medtech Luzern, LausanneTec, SMT Genève, Platinn, BioAlps	Support the creation of networks between actors and the development of a medical market	Networking support between medical local actors legitimizing global and technical opinion

Source: Author's own work.

[a]To provide some degree of anonymity for the respondents, only the names of actors that agreed to be cited are used in this paper.

reasons of the actual evolution of local medtech start-ups and (3) the socio-economic and spatial relations of actors involved in this evolution (e.g. MNEs, investors and associations).

As our inquiry is based on qualitative methods, our aim was not to consider it as representative, but to allow us to propose a description of a different logic characterizing the cluster. Following Flyvbjerg (2006), the main objective was not to generalize from our case study on medical technologies but to propose a key to understand firm creation from the actors' perspective. In fact, this case-study research enabled us to better understand the processes at play within local innovative businesses, to comprehend the way in which these businesses have evolved, to uncover the relationships forged between these businesses and the multinationals and to gain a better insight into the impact that intermediary actors have on these processes.

From Regional Incubation to Multinational Buyout

While technical skills are straightforward enough to come by through local research institutes, in the case of medtech in Western Switzerland it is clear that start-ups struggle to find the local financial backing required for their research and development needs. According to Crevoisier (1997), new firms are often initially supported through local finance. However, with copious red tape and high production costs, medtech start-ups are increasingly dependent upon multi-local financing for their development. Innovative businesses' development is, therefore, no longer solely linked to the region's capacity to provide local investment (bank loans, public authority support, etc.), but to their capacity to attract the interest and support of the major groups organized at the global level. This model is overturning the way local innovation systems work, as most start-ups are generally founded by entrepreneurs looking for the business to be sold on the market at a profit.

On the basis of our case study, we have identified three typical phases in the development of a medtech start-up (Figure 4).

Characterized primarily by a general lack of their own resources and by a great deal of uncertainty, the first phase of a medtech start-up's life cycle is all about ideation. In an uncertain environment, securing financial capital for innovation often becomes extremely difficult and requires the mobilization of external resources using alternative network-based strategies. In Western Switzerland, the creation of these specific start-ups and the

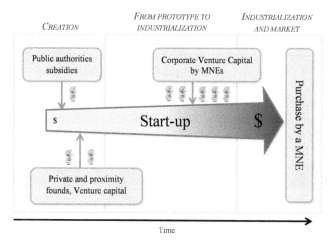

Figure 4. The three phases of Medtech start-up financing.
Source: Author's own work.

design of the first prototypes are modestly financed and supported by government subsidies (Thierstein & Wilhelm, 2001) as well as through local capital (Crevoisier, 1997) often obtained from the entrepreneurs' own social network (e.g. family, friends and contacts). Often the start-up process for innovative businesses involves the financial support of private investors, primarily venture capitalists.

Rarely falling within the ambit of the entrepreneur's social network, their financing of development prototypes is done with a view to eventually making a profit. A great example of this kind of local investment refers to a local start-up that created a special glaucoma-detection contact lens. In fact, the statement of its entrepreneur exposes: "At the beginning, we found our principal investments locally, thanks to friends and family, especially for the creation of a prototype of our contact lens."[5] This statement illustrates that the financial resources necessary to develop the first product prototype came principally through the business owner's own social network and certain public financing initiatives aimed at the incubation and start-up of so-called pre-competitive projects.

The second phase of the start-up's life cycle covers the transition from prototype to industrial production. In contrast to the start-up's initial phase, the phase covering the transition from prototyping to industrial production of products requires much greater sums of money, which are harder to source locally from small or medium-sized investors. The production costs and the cost of the various requisite medical device certifications often necessitate considerable investments, that is, in the region of 20–30 million Swiss Francs.[6] Only multinationals are able to provide this kind of liquidity for producing and certifying new devices. These investments are very often based on corporate venture capital (Chesbrough, 2002; Garel & Jumel, 2005), whereby multinationals take a shareholding in promising companies (Ben Hadj Youssef, 2006). The diabetes treatment device made by a local start-up is a good case in point as the statement of a chief development officer confirms: "The certification process and first phase of industrial production were supported by an American multinational based in the region which now plays a decision-making role in our firm."

The third phase primarily involves the market launch of the medical device and the buyout of the start-up by a multinational. According to Narula and Santangelo's thesis (2009), medical multinationals based in the region achieve innovation not only through the skills within their organization but also through outsourcing and the skills of their external partners, be they research laboratories or businesses. They maintain ongoing relationships, both financial and technological, with local entrepreneurial networks and start-ups in order to benefit from their output and with a view to possibly buying out the start-up, internalizing its product, its production and its specialist workforce. For example, a young business based in Western Switzerland, specializing in the development, production and marketing of implantable medical devices and accessories, was bought out by a big American group in 1994 and was incorporated as a new business within the family of this large group. The statement of a collaborator of this medical company exemplified this issue: "It was 1994, a big American Group, focused on the production of a similar product as ours, showed an interest for our company and we finally decided to sell it them." Similarly, the sale of medtech start-ups to quoted groups may also depend on the entrepreneur's willingness, as illustrated by the statement of a local entrepreneur:

When I established my new enterprise and started conceptualizing a prototype of the product, I was sure that I would sell my start-up to a big group in the near future in order to have enough money to then start a new business.

Looking at these three stages of evolution, we note two main issues. Firstly, start-ups develop through both public and private local capital based on a relationship of trust between actors (Crevoisier, 1997). Unlike the Silicon Valley ideal-type (based on an influx of venture capital enabling the rapid creation of start-ups) (Comtesse, 2013, p. 14), venture capitalists play only a minor role in the setting up of new medical businesses in Western Switzerland. According to the Garel and Jumel approach (2005), despite their minimal involvement in the creation of medtech start-ups, more substantial investments come in the form of corporate venture capital bestowed by large stock-market-listed companies. Although these investments support start-ups through the process of certification and the initial manufacturing of products, they also enable the multinational to gain easy access to new technologies, to improve internal research and development through the applications developed by the start-up, to identify new markets and, indeed, to create a profit (Chesbrough, 2002; Ben Hadj Youssef, 2006).

Secondly, these processes indicate a radical change to the industry's traditional processes of innovation and development. The typical view of the innovative entrepreneur is that of a person setting up a business with the aim of creating endogenous growth through the sale of his or her product (Lebret, 2007). The product is the item to be commoditized and the start-up is the means by which it is invented and put on the market. Therefore, the case of the medical industry in Western Switzerland evidences not only the buyout of start-ups by multinationals but also a desire on the part of entrepreneurs to create a business with the aim of selling it on to a large group within the short to medium term. Although the buyout of start-ups by listed companies is hardly unusual, the commonly expressed desire of entrepreneurs to sell on their business is a more recent phenomenon. In such cases, the medical product becomes more a means of increasing a company's value in the start-up market. The proliferation over the last 10 years of prizes and quality kitemarks being awarded to start-ups rather than to specific products is a good illustration of this turnaround. For example, a local start-up received in 2013 a prize from public authorities considering it as the third best innovative start-up in Switzerland. According to its entrepreneur, this prize "allowed the start-up to find more easily new investors for the creation of the firm and new buyers".

The Construction of the Medtech Start-up Market

Traditionally, start-ups are actors dedicated to creating new products for exploitation on the market which should enable them to develop into SMEs (Lebret, 2007). Our case study shows that start-ups do not concentrate solely on the creation of medical devices to be sold on the market. They also increase their intrinsic market value, thanks to the support of intermediary actors (e.g. opinion leaders) and the development of profitable products. On the one hand, this increase in value enables private investors (and particularly the initial venture capitalists) to make a profit at the first stage of the products' sale and at the final stage of the start-up's sale to the large stock-market-listed groups. On the other hand, this increase in value enables the entrepreneur to make more money, thanks to the buyout of the start-up by a multinational, which often already has a stake in the company through corporate venture terms (Garel & Jumel, 2005).

Generally speaking, medtech start-ups are being less and less viewed as nascent productive companies and more and more as socio-technical devices designed to be marketed. In other words, the medical product developed by a start-up becomes just one of various identifying features of a marketing concept which is sold in the form of a complex entrepreneurial project.

In a process of perpetual reinvestment, money made on the sale of the start-up is often reinjected into the system by the entrepreneur to set up a new one. Developed through a regional business incubator, a local start-up is involved in the development of neurostimulation technologies enabling improved therapies for neurological diseases. This start-up is a good example of the described phenomenon, as the statement of the main contractor confirms: "During the creation phase of my start-ups, my objective was clearly defined; I wanted to create a new medical application and a new enterprise potentially attractive for investors and specific multinational enterprises interested in a firm purchase." This innovative business was recently bought out by an American multinational, wanting to apply the concept developed by the start-up to its own products. Consequently, the medical device is no longer just considered a panacea through which to develop the business and to expand to become leaders within their sector; rather, the product is seen as the medium through which the business's image is to be promulgated, to attract investors and ultimately sell the start-up.

In our particular case, the construction of the medtech start-up market in Western Switzerland occurs through the intermediation of "legitimating third parties" whose power is enacted in specific events and forums. While an entrepreneur's network of contacts and their friends and family may provide the necessary cognitive and financial resources to get a start-up going (Crevoisier, 1997; Grossetti, 2006), intermediary actors help boost the value and legitimacy of an innovative business in the eyes of investors and market product distributors. When seeking financing, a market for their products and the requisite certification for their devices, entrepreneurs call upon opinion leaders, key figures in the medtech sector. Crucial players in the dissemination of new products (Van Eck et al., 2011), they are capable of influencing the opinions, attitudes, motivations and behaviour of others, and define themselves as "people in a social network who, in the diffusion of product and technologies, have greatest influence on their acknowledgment or adoption by other people" (Cho et al., 2012, p. 97). Largely made up of specialist doctors within the medical field of the product in question, opinion leaders provide legitimacy and credibility for both the start-up and the medical device, firstly in the eyes of investors and secondly, of distributors and their customers (Figure 5). Without this specific support, entrepreneurs would be hard-pressed to mobilize the capital required to develop their products or to raise the interest of investors and distributors in their medical applications.

In opposition to Florida and Kenney's arguments (1988), investors' evaluations of start-ups and their products are not based purely on the criteria of originality, patents or the presence of competitive businesses on the market. On the other hand, investors' valuations are based primarily on quality considerations (in relation to ratings agency certifications) as well as the credibility afforded them by opinion leaders and a product's potential range of applications. Similarly, the legitimization of these products by opinion leaders occurs in specific locations, notably business platforms (Cooke et al., 2011; Gawer, 2011), represented by trade fairs, specialist shows or medical conferences. These privileged

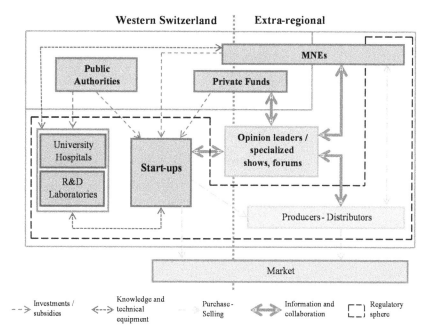

Figure 5. Medtech start-ups in the local environment and the role of legitimating third parties. *Source:* Author's own work.

meeting places shape relations between actors by, on the one hand, enabling entrepreneurs to present their own business to specialists within the field, and, on the other hand, conferring both technical and symbolic value upon the products through the support of opinion leaders.

Discussion: Re-conceptualizing Start-ups in Regional Innovation

Based on the case of the medical industry in Western Switzerland, and more specifically start-ups in the region, we have discerned three key issues (Table 2): (1) the development trajectory of start-ups and the resources mobilized, (2) the modalities of their market evaluation and valorization and (3) the spatial organization underlying their evolution.

Firstly, medtech start-ups are primarily born of entrepreneurial instinct, which follows on from an entrepreneur's higher education. Unlike the traditional creation of innovative companies by the "intrapreneuriat" (Hulsink & Manuel, 2006; Hatchuel *et al.*, 2009), the creation of innovative medtech businesses is less dependent on the entrepreneur's having prior experience in a relevant business. In fact, it is often the result of an individual attempting to respond to real-life problems that he or she has encountered, and the desire to ultimately make a profit when it is sold to a large group. In this situation, the requisite technologies are often to be found in the region, thanks to its numerous research laboratories. By analogy with Crevoisier (1997), when creating start-ups, entrepreneurs primarily rely upon their own personal network of contacts for financial resources. However, in the product certification and manufacturing phase, it is corporate venture capital (Garel & Jumel, 2005) which is most often relied upon.

Table 2. Two contrasted start-up approaches

	Start-up as nascent enterprise/industry	Start-up as products
Entrepreneurship/ emergence	Entrepreneurial project of a productive business	Market solution and entrepreneurial concept incubated after leaving higher education.
Technology	To develop and exploit within production	Made available as a concrete prototype and entrepreneurial concept
Product	Market commodity	Socio-technical component of an entrepreneurial concept
Objective of entrepreneur	Development of a new product (exploitation over the long term)	Selling the start-up to a MNE during the emergence phase (added value on equity)
Investments	Proximity capital and traditional bank loans	Public and private proximity capital and corporate venture investments
Start-up evaluation	In the product market	Legitimacy of the entrepreneurial concept by intermediaries (credibility, trust of opinion leaders, etc.) and financing actors
Territoriality of relations (temporal evolution)	1 Local	Local and multi-local
	2 Supplements coming from elsewhere	Multi-local (selective anchoring)

Source: Author's own work.

Secondly, the territorial aspects of the innovation processes studied and the medtech start-ups in Western Switzerland indicate that medical devices are now created and developed using resources from both local sources (local capital and technologies) and multi-local sources (venture capital and corporate venture capital). Interactions between regional actors are based on local relationships of trust, similar to those described in the approaches to territorial innovation models (Moulaert & Sekia, 2003). These actors develop networks which enable them to combine regional competencies and to create productive synergies (Scott, 2006). However, innovative businesses are increasingly part of networks which transcend regional boundaries. Medtech start-ups both need and benefit from multi-local networks in order to attract the requisite financial resources to develop their products and to interact with key market players such as distributors and opinion leaders. Thus, the local anchoring and subsequent development of the innovative business are quite weak following the start-up's acquisition by a multinational, which often relocates it outside of the region.

Thirdly, the ultimate aim of medtech start-ups and those who start them is no longer to develop a new business producing medical devices over the long term, but rather to create a socio-technical concept to sell on to a large group in the medical industry within the short to medium term. For innovative businesses to attract the attention of these large groups, start-ups require the support of key intermediary figures: opinion leaders. As highlighted by Van Eck *et al.* (2011), these actors confer legitimacy upon and create confidence around start-ups and their products, enabling them to attract the necessary resources to set up a business and create products. In this particular situation, the start-up's value is based not only on technical factors but also on the legitimacy and symbolic value conferred upon it by intermediary actors.

Conclusion: What Life Cycle Approach to Territorial Competitiveness?

Characterized by much diversified firms, technologies and products, medtech activities in Western Switzerland are primary related through their common market orientation. Highly regulated by international technical and safety norms and organized around large strategic players (e.g. hospitals or large medical equipment suppliers), entering such a market is particularly difficult for new comers. Building up their own production tools and distribution channels often requires unaffordable investments for start-ups.

In this context, strategic partnerships or mergers with established multinational companies are usually seen as the most pragmatic—if not the only—way to pursue their industrial development. Consequently, start-ups tend to be conceived from their creation as products commoditized and qualified by various socio-technical devices in market (e.g. certificates, awards or opinion leaders). If such a phenomenon is particularly enhanced in the context of medtech activities, more general considerations and concerns for territorial innovation can be drawn out of this specific case.

If the dual dimension of start-ups, either as "nascent enterprise/industry" or as "product", has always been recognized (Kenny & Von Burg, 1999), territorial innovation models and policy (best) practices traditionally tend to associate their emergence and growth with the development of local industries, either as industrial pioneers or as innovative spin-offs embedded in a regional production system. This vision reflects a spatial division of labour "pushed" by innovators and investors originating from developed countries. This also postulates a limited mobility of production factors (e.g. firms, technologies and workers) in the industrial growing phase of regional innovation.

Adopting such a traditional approach leads to interpreting the case of medtech activities in Western Switzerland as the emergence of a new cluster meant to develop and create new competitive enterprises and industrial boundaries, to provide new regional growth of revenue and employment in the region. In other words and by analogy to the canonical model, this could be viewed as the emergence of a Swiss "Medtech Valley".

However, such an interpretation could fall short of a pertinent analysis when considering future challenges for regional medtech activities. A greater emphasis on start-ups as "products" raises new avenues for research and policy approaches to territorial innovation and competitiveness. Three fundamental open questions seem to us crucial to be asked in this regard (Figure 6).

Firstly, "will regional innovation systems be confined to short-run project life cycles of emergence and reemergence?" In the medtech case examined in the article, entrepreneurial projects seem to consist more and more in combining existing knowledge (science, technology and culture based) in a business idea and advertising it through a start-up concept meant to generate itself a profit. Such projects mirror a knowledge-based economy marked by permanent innovation where the constant renewal of entrepreneurial projects becomes the key factor of competitiveness. In this view, more stable phases of industrialization tend to be shaded by perpetual sequences of emergence and reemergence. Beyond a life cycle approach, future research avenues and policy issues will have to be able to identify the industrial dynamics of growth induced, or not, by this "project

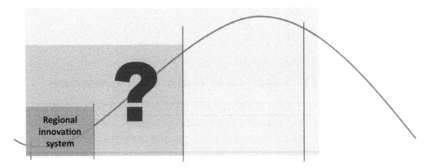

Figure 6. What new life cycles of territorial growth?
Source: Author's own work.

ecology" (Grabher, 2002b) and to understand how they contribute, or not, to broader regional innovation systems.

Pursuing a same line of reflection, a second provocative question arises: "will commoditized entrepreneurial projects become the new revenue model of lead innovative regions?" Regional competitiveness is usually conceived as the capacity to produce innovative goods or activities and to export them. Sold as products, entrepreneurial projects become themselves commoditized revenues for innovative regions. Within global production and financial networks, innovators and investors are, nowadays, not only situated in the traditionally called "developed countries". Highly mobile, knowledge and financial resources circulate and anchors across territories, pulled by foreign direct investments chasing most attractive innovations and entrepreneurial projects. Purchased in the early phase of emergence, innovations can be more easily relocated according to the corporate strategies of multinational companies. Depicting today's territorial revenue models should be at the core of future research and policies agenda to understand how regional wealth is actually generated. Traditional export-base models of growth will certainly have reconsidered in such enterprise.

Finally, these two fundamental questions underlie the general exploratory question: "how current regional innovation systems will be able to position themselves within global networks to attract and anchor knowledge and financial resources?" As pointed out by the case of Swiss medtech start-ups, being competitive today is not only being innovative. Policies of knowledge creation and transfer are for sure influent factors of regional innovation. However, as illustrated earlier, local innovation is also about being able to advertise local entrepreneurship to global players and investors. It is also about thinking how such projects and the revenue they generate will stimulate the renewal of future projects as well as develop within broader industrial growth.

Acknowledgements

The authors thank the *Swiss National Science Foundation* and the *European Science Foundation* for their support to the research entitled "Cluster Emergence, Renewal and Transition in Switzerland: Evidence from Cleantech, Medtech and the Watch Industry." They also thank Max-Peter Menzel and the two anonymous reviewers for their precious comments on earlier versions of this paper. The responsibility for the content of this contribution remains entirely ours.

Notes

1. Data taken from a survey into the Swiss medical technologies industry in 2012, published by Medtech Switzerland (an initiative of the Swiss government) in collaboration with Fasmed and Medical Cluster (Medtech Switzerland, 2012).
2. Research financed by the *Fond National Suisse de Recherche Scientifique* entitled "Cluster Emergence, Renewal and Transition in Switzerland: Evidence from Cleantech, Medtech and the Watch Industry."
3. According to Corbin and Strauss (2008), the saturation concerns reaching the point where "the new" does not necessarily add anything to the story, model, theory or framework.
4. MAXQDA was used as a Computer-Assisted Program to analyse our qualitative data.
5. All statements in this paper have been translated from French by the authors.
6. In the region of 16–24 million Euros.

References

Asheim, B. & Coenen, L. (2005) Knowledge bases and regional innovation systems: Comparing Nordic clusters, *Research Policy*, 34(8), pp. 1173–1190.

Asheim, B., Boschma, R. & Cooke, P. (2011) Constructing regional advantage: Platform policies based on related variety and differentiated knowledge bases, *Regional Studies*, 45(7), pp. 893–904.

Audia, P. G. & Rider, C. I. (2005) A garage and an idea: What more does an entrepreneur need?, *California Management Review*, 48(1), pp. 6–28.

Bathelt, H. & Glückler, J. (2005) Resources in economic geography: From substantive concepts towards a relational perspective, *Environment and Planning A*, 37(9), pp. 1545–1563.

Bathelt, H., Malmberg, A. & Maskell, P. (2004) Clusters and knowledge: Local buzz, global pipelines and the process of knowledge creation, *Progress in Human Geography*, 28(1), pp. 31–56.

Ben Hadj, Youssef, A. (2006) Le capital risque: que vont faire les grands groupes dans les start-ups?, *Gérer et Comprendre*, 84, pp. 34–43.

Boschma, R. A. & Lambooy, J. G. (1999) Evolutionary economics and economic geography, *Journal of Evolutionary Economics*, 9, pp. 411–429.

Brenner, T. & Schlump, C. (2011) Policy measures and their effects in the different phases of the cluster life cycle, *Regional Studies*, 45(10), pp. 1363–1386.

Carrincazeaux, C. & Coris, M. (2011) Proximity and innovation, in: P. Cooke, B. Asheim, R. Boschma, R. Martin, D. Schwartz & F. Tödling (Eds) *Handbook of Regional Innovation and Growth*, pp. 269–281 (Cheltenham: Edward Elgar).

Chataway, J. & Wield, D. (2000) Industrialization, innovation and development: What does knowledge management change?, *Journal of International Development*, 12(6), pp. 803–824.

Chesbrough, H. (2002) Making sense of corporate venture capital, *Harvard Business Review*, 80(3), pp. 4–11.

Cho, Y., Hwang, J. & Lee, D. (2012) Identification of effective opinion leaders in the diffusion of technological innovation: A social network approach, *Technological Forecasting and Social Change*, 79(1), pp. 97–106.

Coe, N. M., Lai, K. P. Y. & Wójcik, D. (2014) Integrating finance into global production networks, *Regional Studies*, 48(5), pp. 761–777.

Coffey, W. J. & Polèse, M. (1984) The concept of local development: A stages model of endogenous regional growth, *Papers in Regional Science*, 55(1), pp. 1–12.

Comtesse, X. (2013) *La santé de l'innovation suisse: pistes pour son renforcement* (Genève: Avenir Suisse).

Cooke, P. (2001) Regional innovation systems, clusters, and the knowledge economy, *Industrial and Corporate Change*, 10(4), pp. 945–974.

Cooke, P. (2008) Regional innovation systems, clean technology & Jacobian cluster-platform policies, *Regional Science Policy & Practice*, 1(1), pp. 23–45.

Cooke, P. & Martin, R. (Eds) (2006) *Clusters & Regional Development* (Hampshire: Routledge).

Cooke, P., Asheim, B., Boschma, R., Martin, R., Schwartz, D. & Tödling, F. (2011a) *Handbook of Regional Innovation and Growth* (Cheltenham: Edward Elgar).

Cooke, P., Asheim, B., Boschma, R., Martin, R., Schwartz, D. & Tödling, F. (2011b) Introduction to the handbook of regional innovation and growth, in: P. Cooke, B. Asheim, R. Boschma, R. Martin, D. Schwartz, & F. Tödling (Eds) *Handbook of Regional Innovation and Growth*, pp. 1–23 (Cheltenham: Edward Elgar).

Cooke, P., De Laurentis, C., Macneill, S. & Collinge, C. (2011) *Platforms of Innovation: Dynamics of New Industrial Knowledge Flows* (Cheltenham: Edward Elgar Publishing).

Corbin, J. & Strauss, K. (Eds) (2008) *Basics of Qualitative Research* (London: Sage).

Corpataux, J., Crevoisier, O. & Theurillat, T. (2009) The expansion of the finance industry and its impact on the economy: A territorial approach based on Swiss pension funds, *Economic Geography*, 85(3), pp. 313–334.

Crevoisier, O. (1997) Financing regional endogenous development: The role of proximity capital at the age of globalization, *European Planning Studies*, 5(3), pp. 407–415.

Crevoisier, O. & Jeannerat, H. (2009) Territorial knowledge dynamics: From the proximity paradigm to multi-location milieus, *European Planning Studies*, 17(8), pp. 1223–1241.

Delgado, M., Porter, M. E. & Stern, S. (2010) Clusters and entrepreneurship, *Journal of Economic Geography*, 10(4), pp. 495–518.

Doloreux, D. (2002) What we should know about regional systems of innovation, *Technology in Society*, 24(3), pp. 243–263.

Etzkowitz, H. (2006) The new visible hand: An assisted linear model of science and innovation policy, *Science and Public Policy*, 33(5), pp. 310–320.

Etzkowitz, H. & Leydesdorff, L. (2000) The dynamics of innovation: From national system and "Mode 2" to a triple-helix of university-industry-government relations, *Research Policy*, 29(2), pp. 109–123.

Feldman, M. P. (2001) The entrepreneurial event revisited: Firm formation in a regional context, *Industrial and Corporate Change*, 10(4), pp. 861–891.

Feldman, M. P., Francis, J. & Bercovitz, J. (2005) Creating a cluster while building a firm: Entrepreneurs and the formation of industrial clusters, *Regional Studies*, 39(1), pp. 129–141.

Ferrary, M. (2008) Les capital-risqueurs comme "transiteurs" de l'innovation dans la silicon valley, *Revue Française de Gestion*, 10(190), pp. 179–196.

Florida, R. L. & Kenney, M. (1988) Venture capital, high technology and regional development, *Regional Studies*, 22(1), pp. 33–48.

Flyvbjerg, B. (2006) Five misunderstandings about case-study research, *Qualitative Inquiry*, 12(2), pp. 219–245.

Frenken, K. & Boschma, R. (2007) A theoretical framework for evolutionary economic geography: Industrial dynamics and urban growth as a branching process, *Journal of Economic Geography*, 7(5), pp. 635–649.

Fritschi, H. (2006) Medizinaltechnologie: Unter riesen, *Bilanz*, January 31, pp. 53–55.

Garel, G. & Jumel, S. (2005) Les grands groupes et l'innovation: définitions et enjeux du corporate venture, *Finance, Contrôle, Stratégie*, 8(4), pp. 33–61.

Gawer, A. (Ed.) (2011) *Platforms, Markets and Innovation* (Cheltenham: Edward Elgar).

Gelijns, A. C. & Thier, S. O. (2002) Medical innovation and institutional interdependence: Rethinking university-industry connections, *The Journal of the American Medical Association*, 287(1), pp. 72–77.

Gerring, J. (2007) *Case Study Research. Principles and Practices* (New York: Cambridge University Press).

Gibbons, M., Limoges, C., Nowotny, H., Schwartzman, S., Scott, P. & Trow, M. (1994) *The New Production of Knowledge. The Dynamics of Science and Research in Contemporary Societies* (London: Sage).

Glaeser, E. L. & Kerr, W. R. (2009) Local industrial conditions and entrepreneurship: How much of the spatial distribution can we explain?, *Journal of Economics and Management Strategy*, 18(3), pp. 623–663.

Grabher, G. (1993) The weakness of strong ties: The lock-in of regional development in the Ruhr area, in: G. Grabher (Ed.) *The Embedded Firm: On the Socioeconomics of Industrial Networks*, pp. 255–277 (London: Routledge).

Grabher, G. (2002a) Cool projects, boring institutions: Temporary collaboration in social context, *Regional Studies*, 36(3), pp. 205–214.

Grabher, G. (2002b) The project ecology of advertising: Tasks, talents and teams, *Regional Studies*, 36(3), pp. 245–262.

Grbich, C. (Ed.) (2012) *Qualitative Data Analysis: An Introduction* (London: Sage).

Grossetti, M. (2006) Réseaux sociaux et ressources de médiation dans l'activité économique, *Sciences de la Société*, 73, pp. 83–103.

Guillemette, F. & Luckerhoff, J. (2009) L'induction en méthodologie de la théorisation enracinée (MTE), *Recherches Qualitatives*, 2, pp. 4–21.

Hatchuel, A., Garel, G., Le Masson, P. & Weil, B. (2009) L'intrapreneuriat, compétence ou symptôme ?. Vers de nouvelles organisations de l'innovation, *Revue Française de Gestion*, 195, pp. 159–174.

Hilhorst, J. G. M. (1998) Industrialization and local/regional development revisited, *Development and Change*, 29(1), pp. 1–26.

Hulsink, W. & Manuel, D. (2006) Venturing into the entrepreneurial unknown: On entrepreneurship in the high-tech industries. Paper prepared for the 14th Annual High Technology Small Firms Conference, 11–13 Mai, Twente, the Netherlands. Available at http://proceedings.utwente.nl/144/1/Hulsink,W._cs_paper1.pdf (accessed 11 June 2014).

Isaksen, A. (2011) Cluster evolution, in: P. Cooke, B. Asheim, R. Boschma, R. Martin, D. Schwartz, & F. Tödling (Eds) *Handbook of Regional Innovation and Growth*, pp. 293–302 (Cheltenham: Edward Elgar).

Julien, P.-A. & Marchesnay, M. (2011) *L'Entrepreneuriat* (Paris: Economica).

Kenney, M. & Von Burg, U. (1999) Technology, entrepreneurship and path dependence: Industrial clustering in Silicon Valley and route 128, *Industrial and Corporate Change*, 8(1), pp. 67–103.

Kiese, M. & Schätzl, L. (Eds) (2008) *Cluster und Regionalentwicklung: Theorie, Beratung und praktische Umsetzung* (Dortmund: Rohn).

Klepper, S. (1997) Industry life cycles, *Industrial and Corporate Change*, 6(1), pp. 145–182.

Klöpper, C. & Haisch, T. (2008) Evolution de l'industrie biotech et medtech suisse et influence de l'industrie pharmaceutique sur le système d'innovation, *Revue Géographique de l'Est* [online], 48(3–4). Available at http://rge.revues.org/1694 (accessed 11 December 2013).

Lambooy, J. (2005) Innovation and knowledge: Theory and regional policy, *European Planning Studies*, 13(8), pp. 1137–1152.

Lebret, H. (2007) *Start-up: ce que nous pouvons encore apprendre de la Silicon Valley* (Lausanne: CreateSpace Independent Publishing Platform).

Maillat, D. (1995) Territorial dynamic, innovative milieus and regional policy, *Entrepreneurship and Regional Development*, 7(2), pp. 157–165.

Markusen, A. (1985) *Profit Cycles, Oligopoly and Regional Development* (Cambridge: The MIT Press).

Marshall, A. (1890) *Principles of Economics* (London: Macmillan).

Martin, R. & Sunley, P. (2003) Deconstructing clusters: Chaotic concept or policy panacea?, *Journal of Economic Geography*, 3(1), pp. 5–35.

Martin, R. & Sunley, P. (2011) Conceptualizing cluster evolution: Beyond the life cycle model?, *Regional Studies*, 45(10), pp. 1299–1318.

Medtech Switzerland (2012) *Swiss Medtech Report* (Bern: Medtech Switzerland).

Menzel, M. P. & Fornahl, D. (2010) Cluster life cycles – dimensions and rationales of cluster evolution, *Industrial and Corporate Change*, 19(1), pp. 205–238.

Merriam, S. B. (1998) *Qualitative Research and Case Study Applications in Education* (San Francisco, CA: Jossey-Bass Publishers).

Moulaert, F. & Sekia, F. (2003) Territorial innovation models: A critical survey, *Regional Studies*, 37(3), pp. 289–302.

Nagy Hesse-Biber, S. & Leavy, P. (2011) *The Practice of Qualitative Research*, 2nd ed. (London: Sage).

Narula, R. & Santangelo, G. D. (2009) Location, collocation and R&D alliances in the European ICT industry, *Research Policy*, 38(2), pp. 393–403.

Penrose, E. T. (1952) Biological analogies in the theory of the firm, *The American Economic Review*, 42(5), pp. 804–819.

Polèse, M. & Shearmur, R. (2009) *Économie Urbaine et Régionale*, 3ème ed. (Paris: Economica).

Porter, M. E. (1998) Clusters and the new economics of competition, *Harvard Business Review*, 76(6), pp. 77–90.

Potter, A. & Watts, D. J. (2011) Evolutionary agglomeration theory: Increasing returns, diminishing returns and the industry life cycle, *Journal of Economic Geography*, 11(3), pp. 417–455.

Qvortrup, L. (2006) The new knowledge regions: From simple to complex innovation theory, in: P. Cooke, & A. Piccaluga (Eds) *Regional Development in the Knowledge Economy*, pp. 246–271 (Abingdon: Routledge).

Rocha, H. O. (2004) Entrepreneurship and development: The role of clusters, *Small Business Economics*, 23(5), pp. 363–400.

Saxenian, A. (1990) Networks and the resurgence of Silicon Valley, *California Management Review*, 33(1), pp. 89–112.

Saxenian, A. (1991) The origins and dynamics of production networks in Silicon Valley, *Research Policy*, 20(5), pp. 423–437.

Saxenian, A. (1994) *Regional Advantage: Culture and Competition in Silicon Valley and Route 128* (Cambridge, MA: Harvard University Press).

Schmitz, H. (1999) Collective efficiency and increasing returns, *Cambridge Journal of Economics*, 23(4), pp. 465–483.

Schmitz, H. & Nadvi, K. (1999) Clustering and industrialization: Introduction, *World Development*, 27(9), pp. 1503–1514.

Schumpeter, J. A. (1935) *Théorie de l'Evolution Economique* (Paris: Dalloz).

Schumpeter, J. A. (1939) *Business Cycles: A Theoretical, Historical and Statistical Analysis of the Capitalist Process* (New York: McGraw Hill Book Co).

Scott, A. J. (1986) Industrialization and urbanization: A geographical agenda, *Annals of the Association of American Geographers*, 76(1), pp. 25–37.

Scott, A. J. (1988) Flexible production systems and regional development: The rise of new industrial spaces in North America and Western Europe*, *International Journal of Urban and Regional Research*, 12(2), pp. 171–186.

Scott, A. J. (2006) Entrepreneurship, innovation and industrial development, *Small Business Economics*, 26(1), pp. 1–24.

Scott, A. J. & Storper, M. (Eds) (1992) *Pathways to Industrialization and Regional Development* (London: Routledge).

Silverman, D. (Ed.) (2010) *Qualitative Research* (London: Sage).

Simmie, J. (2005) Innovation and space: A critical review of the literature, *Regional Studies*, 39(6), pp. 789–804.

Stake, R. E. (2005) Qualitative case studies, in: N. K. Denzin & Y. S. Lincoln (Eds) *Qualitative Research*, pp. 443–465 (London: Sage).

Stam, E. (2007) Don't leave: Locational behavior of entrepreneurial firms, *Economic Geography*, 83(1), pp. 27–50.

Stam, E. (2010) Entrepreneurship, evolution and geography, in: R. Boschma & R. Martin (Eds) *Handbook on Evolutionary Economic Geography*, pp. 139–161 (Cheltenham: Edward Elgar).

Sternberg, R. & Litzenberger, T. (2004) Regional clusters in Germany – their geography and their relevance for entrepreneurial activities, *European Planning Studies*, 12(6), pp. 767–791.

Thierstein, A. & Wilhelm, B. (2001) Incubator, technology and innovation centres in Switzerland: Features and policy implications, *Entrepreneurship and Regional Development: An International Journal*, 13(4), pp. 315–331.

Tichy, G. (2011) Innovation, product life cycle and diffusion: Vernon and beyond, in: P. Cooke, B. Asheim, R. Boschma, R. Martin, D. Schwartz, & F. Tödling (Eds) *Handbook of Regional Innovation and Growth*, pp. 67–77 (Cheltenham: Edward Elgar).

Tödtling, F. & Trippl, M. (2005) One size fits all? Towards a differentiated regional innovation policy approach, *Research Policy*, 34(8), pp. 1203–1219.

Trettin, L. & Welter, F. (2011) Challenges for spatially oriented entrepreneurship research, *Entrepreneurship & Regional Development*, 23, pp. 575–602.

Van Eck, P. S., Jager, W. & Leeflang, P. S. H. (2011) Opinion leaders' role in innovation diffusion: A simulation study, *Journal of Product Innovation Management*, 28(2), pp. 187–203.

Vatne, E. (2011) Regional agglomeration and growth: The classical approach, in: P. Cooke, B. Asheim, R. Boschma, R. Martin, D. Schwartz, & F. Tödling (Eds) *Handbook of Regional Innovation and Growth*, pp. 54–65 (Cheltenham: Edward Elgar).

Vernon, R. (1966) International investment and international trade in the product cycle, *The Quarterly Journal of Economics*, 80(2), pp. 190–207.

Weigel, S. (2011) Medical technology's source of innovation, *European Planning Studies*, 19(1), pp. 43–61.

Yin, R. K. (2009) *Case Study Research: Design and Methods* (Los Angeles, CA: Sage).

Adaptation and Change in Creative Clusters: Findings from Vienna's New Media Sector

TANJA SINOZIC & FRANZ TÖDTLING

Institute for Multilevel Governance and Development, Vienna University of Economics and Business, Vienna, Austria

ABSTRACT   *This paper investigates some of the features of technological heterogeneity in the New Media cluster in Vienna and the local and global factors that have shaped territorial learning conditions over time. Technological heterogeneity is given a central role in cluster evolution for the expansion of local capacities and opportunities for change. In this paper, it is argued that technological heterogeneity is an important but insufficient motor for cluster evolution. Rather, what is required are local technological capabilities and learning conditions for the exploitation of technologies for operations and procedures that are relevant for firm and cluster performance, as posited by evolutionary theories of technical change. These perspectives are used to interpret the complex, variegated and partially unpredictable features of technological heterogeneity in the New Media cluster in Vienna, revealing the importance of the capabilities embodied in people and local conditions of managing uncertainty mediated via heterogeneity in products, processes and client needs. For this sector, conditions of technological instability create increasing importance for local learning and networks if clusters are to be propelled more deeply into existent or more radically into novel specializations.*

Introduction

Technological heterogeneity is central to the evolution of clusters (Menzel & Fornahl, 2010). If clusters possess a diverse range of firms, institutions and industries, they are more able to grow and transform into novel clusters under conditions threatening decline (Menzel & Fornahl, 2010). A cluster emerges when the focal points of local firms converge upon one technological area, creating stable cluster networks over time (Menzel & Fornahl, 2010). In contrast, a low level of internal technological diversity can lead a cluster to exhaustion. The view of technological heterogeneity as a driver of cluster growth is still in its emerging stage, so beyond this greater level of generality little insight exists into the precise details of how and why this cluster feature is so important for cluster performance. However, the view that technical change is tightly interwoven in the fabric that underpins growth processes has long been a major topic of enquiry for evolutionary economists (Dosi, 1982, 1988; Freeman, 1982; Rosenberg, 1982). Within this body of literature, some scholars have investigated the types and range of sources of technical change and the importance of technological capabilities for their translation into firm performance (Bell & Pavitt, 1993; Hobday, 1995). This paper seeks to apply the insights from theories of technological capabilities to characterize technological heterogeneity and to investigate the factors that are important for its economic exploitation in creative industries clusters.

Creative industries are growing sectors of modern knowledge economies and have been the focus of much recent research and policy-making (Power & Scott, 2004; Pratt, 2008). Creative industries share the feature that creativity is key in work processes and a central reason for competitiveness of one firm or place over another (Florida, 2002). Although the sectors subsumed under the term vary, authors increasingly describe creative industries as a group of diverse sectors such as software and computer services, scientific research and development, advertising and market research, printing and reproduction of recorded media, motion pictures, video, television production, architecture and engineering, creative arts and entertainment (Florida, 2002; Lazzeretti, 2012).

Creative industries are highly embedded in their localities because of the intertwining of place and culture, the reliance of work processes of geographically dense networks, and high dependence on specialized labour and its flexible recruitment (Scott, 1997). Creative industries tend to be located in urban areas, where large cities act as "hubs" for individuals and firms specialized in creative services (Cooke & Lazzeretti, 2008; Lazzeretti, 2012). Highly qualified labour, diversified knowledge inputs, infrastructure and direct access to clients make cities a fertile ground for creative service firms (Cooke *et al.*, 2008). Studies have shown that supportive conditions can exist in smaller cities and suburban areas (Lazzeretti 2012, Regional Studies Special Issue 2/2013). Trippl *et al.* (2012) found that in Austria, creative industries are concentrated in Vienna and other urban areas, with some patterns of suburbanization and dispersal. Recent longitudinal studies have provided insight into factors affecting change in creative clusters over time such as a social context that has helped a production structure to emerge and decline over a period of 40 years (De Propris & Lazzeretti, 2009). A declining local market and a corresponding increase in competition caused a disintegration of production links and pressures on creativity for rebirth of the Birmingham jewellery cluster (De Propris & Wei, 2007).

Less explored are the aspects of how creative industries adapt and transform qualitatively under changing technological, cultural and local/global conditions and, in turn, how they initiate and support change. Evolutionary approaches to clustering processes have offered a complementary theoretical lens with which to explore how clusters emerge and grow and why they may be persistently successful. Evolutionary theorizing suggests that interactions between firm capabilities, industry life cycles and networks enable sectors to take different routes such as further specialization and emergence of new subsectors (Ter Wal & Boschma, 2011). Cluster change is affected by regional path dependencies such as cultural conditions, technological regimes and production structures that have shaped similar industries in the past (Simmie, 2012). Cluster life cycle theory posits that cluster evolution is derived from local technological heterogeneity that interacts with changing local and global processes to converge upon technological focal points, and thereby initiate developmental shifts along the cluster development path (Menzel & Fornahl, 2010). This paper uses this emerging field of enquiry to supplement existing empirical studies on creative clusters.

In this paper, we continue our previous investigations of the Vienna creative industries where we provided a historical account of quantitative changes in growth and sectoral composition from 1910 to 2012, and relative changes over time between creative industries in Vienna and Austria overall, based on secondary data (Sinozic *et al.*, 2013). This paper investigates some of the main features of technological heterogeneity, and the local, national and global factors that have shaped its relationship to cluster evolutionary processes. Using primary data collected in 25 face-to-face interviews, this paper addresses the following research questions:

- How is technological heterogeneity characterized in the New Media cluster in Vienna?
- How have technological, economic and institutional factors at regional, national and international levels affected the firm's and the cluster's learning conditions over time?

The following section provides the conceptual framework of the paper, synthesizing the main approaches to cluster evolution and highlighting sectoral aspects of creative industries important for cluster learning and innovation processes in previous studies. The third section gives a brief overview of the research context and describes the survey method used. The fourth section presents the results and analysis, focusing on (1) features of technological heterogeneity, areas and focal points in the cluster and (2) barriers and supports at the regional, national and international levels that shape cluster learning conditions and its evolution. The fifth section concludes the paper by summarizing the main findings and their relevance for cluster evolutionary theory.

Conceptual Framework: Cluster Evolution and Features of Creative Industries

Nelson and Winter's (1982) seminal work provided one of the main incentives for evolutionary perspectives on cluster change. According to Nelson and Winter (1982), the economy does not grow in a linear manner because of rational choices made by homogenous agents leading to equilibrium, but is instead characterized by incremental, differentiated, complex, highly interactive, cumulative and partially unpredictable endogenous processes from agents that are boundedly rational and who operate under non-linear disequilibrating conditions. These views have inspired scholars such as Boschma and Frenken (2006) to suggest viewing local industrial transformation as a variation of routines at the spatial level over time. Regional economic performance is continuously enhanced because of favourable selection environments. Boschma and Frenken (2009) suggest a co-evolutionary perspective of local routines and institutions, in particular for understanding the evolution of product innovations and branching into different product areas which may give rise to related but differentiated clusters. Maskell and Malmberg (2007) state that change at the cluster level is constructed by micro-level routines, search processes, memory and history. A central role of interactions between endogenous features of clusters is also clearly argued in Ter Wal and Boschma's (2011) co-evolutionary view. Clusters change because of the dynamics that are created through firm capabilities, industry life cycles and networks (Ter Wal & Boschma, 2011). Pioneering firms initiate the emergence of clusters, but their sustainment of force is dependent upon regional assets such as qualified labour and infrastructures, their position in and structure of networks and the stage of industry life cycle (Ter Wal & Boschma, 2011).

Cluster evolution is also affected by what happens external to the cluster, and during processes of interaction between and via the mediation of cluster elements and extra-cluster phenomena. Regional path dependence theory suggests that aspects of a region condition the development of its industries in one direction or another (Storper, 1997; Boschma & Lambooy, 1999; Martin & Sunley, 2006; Simmie, 2012). Regional features of economy, technology (e.g. paradigms and regimes) and institutions act as constraints or supports to local firm activities. Unpredictability is underlined by Martin and Sunley (2006) who state that different regional paths are created through chance or through historical accidents, and these can also be a trigger for the characteristic developments of a cluster.

A number of perspectives have been put forward to analyse stage-like change processes in clusters, such as the life cycle metaphor (Hassink & Shin, 2005; Bergman, 2008; Menzel & Fornahl, 2010). Most scholarly attention has been given to cluster emergence in which reasons such as accumulation of knowledge and other regional infrastructures such as institutions, industries and culture coincide with industry life cycles and broader scientific advancements to bring new local firms into existence (Braunerhjelm & Feldman, 2006; Fornahl et al., 2010). A cluster is said to enter a second stage in its development (or "expansion" phase, see Bergman, 2008) when relatively large numbers of firms are created and employment rises because knowledge exploration processes have matured into routinized activities that allow firms to raise their performance (Winter, 1984). In this stage, firms also move into the cluster because of low search costs favourable for their innovation processes (Maskell & Kebir, 2005). This stage may align with the growth phase of the industry (Audretsch & Feldman, 1996). Well-established production structures characterize the end of this stage because over time when processes become more predictable and the costs of being located in the cluster outweigh the benefits, firms may find it more profitable to locate elsewhere (Swann, 2002). If regional conditions are conducive for a regime shift, existing skills and routines can be applied to different technological problems and give rise to novel cluster specializations (Swann, 2002; Bergman, 2008).

Scholars are in broad agreement with regards to technological heterogeneity and forms of diversity in being central to cluster evolution (Frenken & Boschma, 2007; Menzel & Fornahl, 2010). Menzel and Fornahl (2010, p. 210) state that "clusters display long-term growth if they are able to maintain diversity". Diversity is found, for example, in firm capabilities and technology areas in a cluster. The cluster moves through its various developmental stages (emergence, growth, maturity/sustainment and decline/renewal) by exploiting the processes of technological convergence and divergence of firm knowledge bases. Important cluster elements are firms and institutions, and the momentum they create through their internal processes and interconnections to change cluster thematic and spatial boundaries. For example, a cluster may emerge as a firms' knowledge base grows and interacts with other local firms expanding the thematic boundary of the cluster increasing the probability for novel technological focal points (Menzel & Fornahl, 2010). The locational specificities of the cluster influence the differentiation between the life cycle of the industry (which is global) and the life cycle of the cluster. If a cluster finds itself in a phase of maturity with insufficient internal diversity to create novel technological focal points for local firms, it is vulnerable to decline (Menzel & Fornahl, 2010). Cluster life cycles have different degrees of change (adaptation, transformation and renewal). Adaptation processes are probable in relatively young clusters that through the adjustment of their learning processes are still able to shift back to a growth phase from a phase of maturity (Menzel & Fornahl, 2010). Transformation and renewal are more fundamental and radical changes are required when a cluster is in a state of decline (Menzel & Fornahl, 2010). Cluster evolution, thus, is underpinned by the interactions between technological heterogeneity in the cluster and changes in cluster learning conditions which are affected by endogenous and exogenous factors and processes, and their interdependencies.

Despite the central role that is given to technological heterogeneity in the theoretical accounts of cluster evolution, it still remains poorly defined and difficult to operationalize and to investigate empirically in the cluster view. For example, by defining technological heterogeneity as diversity in "firms and institutions" (Menzel & Fornahl, 2010), specific evolutionary knowledge-augmenting processes and the changing conditions under which they occur may remain hidden. A more specific definition of aspects of technology which may translate into better performance is provided by evolutionary theorists of technical change. Bell and Pavitt (1997, p. 89) define technological capabilities as "resources needed to generate and manage technical change (including) knowledge, skills, experience, institutional structures and linkages in, between and outside firms". It follows that clusters, if they are to converge upon technological focal points that underlie competitiveness, need to possess technological capabilities and adaptive conditions which allow for exploitation of technical change and the management of technological uncertainty and instability.

The capabilities and local conditions that are required to adapt to technical change differ greatly across industries (Bell & Pavitt, 1993). In industries such as New Media, for which technical change can according to Pavitt's (1984) taxonomy be said to be dominated by "information intensity", and "specialized suppliers", technological heterogeneity is characterized by corporate software systems, design, responding to user needs and process software and product improvements. These technological areas are both focal points of operational activity and sources of technical change that learning efforts are structured by and converge upon.

Creative Industries: Project-Based and Network-Reliant Work Practices

A growing body of work has explored empirically how individual actors (people, firms and institutions) and their groups (such as sectors and communities of practice) perform their activities and innovate. Given the empirical focus of this paper, we summarize the features of work processes in creative industries such as industry characteristics, project-based work, networks and creativity.

Creative industries have been studied for at least two decades because of their increasingly important influence on economic growth, while other sectors face more obvious pressures of globalization. There is disagreement about which typology should be used to group these and related industries. For example, "cultural" industries are industries that have a cultural value separate from their commercial value (such as creative arts, entertainment activities and museums) and tend to be subsidized (Cooke & Lazzeretti, 2008; Pratt, 2008). "Creative" industries, on the other hand, have a cultural component in their product or service that is primarily functional (Pratt, 2008). In Florida's (2002) important work, all these industries are subsumed under the "creative industries" label. Most useful for the evolutionary view taken in this paper is Lazzeretti *et al.*'s (2008) definition which takes into account change over time, with their differentiation between "traditional creative industries" (e.g. printing and reproduction of recorded media, motion picture, video, television, architectural and engineering activities, creative arts, entertainment and museums) and their partial transformation to "non-traditional creative industries" (such as software and computer services, scientific research and development and advertising and market research), the latter including New Media products and services.

The grouping of relatively different sectors under the name creative industries can be traced back to the evolution of UK policy discourse in the 1990s which culminated with the publication of the UK Creative Industries Task Force mapping document (DCMS, 1998) after which operationalization and output measures followed (Pratt, 2008). Lazzeretti (2008) emphasizes the importance of economic resources (such as human capital), actors (such as firms and institutions) and communities (heterogeneous and multicultural) as important for competitiveness in these industries. Pratt (1997) in his study of the creative industries production system states that technology and production are intertwined in these sectors. Moreover, even the high-performing industries such as advertising may be dependent upon the strength of smaller and weaker industries for skills and products.

Creative industries tend to develop and sell products and services organized in projects (DeFillippi & Arthur, 1998; Lorenzen & Frederiksen, 2008). Project-based work has features such as a firm "renting" its human capital and organizing work along temporary structures (DeFillippi & Arthur, 1998). Projects are highly oriented towards clients, who influence the interactions between creative organizations (such as advertising) (Grabher, 2001). An important driver of inter-organizational interactions is technological diversity within projects, such as, for example, in advertising firms, client needs may not only be advertising but also marketing and communication strategies (Grabher, 2001).

Projects in creative industries tend to be based upon, and over time create, stable communities and networks between individuals and organizations in the region (Sydow & Staber, 2002). Over time, individuals who repeatedly worked together in film production created stability in relationships through routines and practices (Sydow & Staber, 2002). Moreover, what is learned from previous collaborations can then be used to "formulate common strategies for future projects" (Sydow & Staber, 2002, p. 219). Interactions in projects requiring creativity tend to be ambiguously structured because requirements change over time, and it is impossible to predict where the next idea or problem will arise (Hatch, 1999). Continuous learning, experimentation and the questioning of assumptions are features of the creative process (Barrett, 1998) which underlie capabilities in these industries.

Sectoral Background and Method

Vienna's creative industries have a rich history. Resch (2008) did a comprehensive analysis of growth and change in the creative industries in Vienna using Austrian national census statistics from 1910, 1951 and 2001, of which several important features are briefly reviewed here.[1] In 1910, the creative industries in Vienna (composed at the time of traditional creative industries, such as architecture, audio-visuals, visual arts and the arts market, performance art, print and publishing, music, museums and libraries) employed around 200,500 persons. Between 1910 and 1951, Vienna underwent major change in its position in Europe, causing a decline in sectors such as graphics, fashion and design and museums and libraries. During the same period, spurred on by new technology and growing demand, the audio-visuals and music sectors grew. Between 1951 and 2001, some sectors went through dramatic growth phases (especially architecture, museums, libraries, advertising, architecture and audio-visuals). Graphics, fashion, design, print, publishing and music all declined during this period, and never really recovered. This period marked the emergence of the global information and communication technologies (ICTs) industry, and the start of New Media. Between 2000 and 2010, the sectors that have converged to form the New Media cluster in Vienna (such as advertising, software and computer services) have grown by approximately by 40%, the most dramatic growth of all creative industries during that period. These sectors have also been the major focus of government subsidies during this period.

For the operationalization of factors and processes that characterize and condition local technological heterogeneity in the New Media cluster, we relied upon existing contextualized empirical studies on technical change and technological capabilities in sectors (Hobday, 1995) and on studies investigating the emergence and growth of clusters and innovation in clusters (Braunerhjelm & Feldman, 2006). While cluster growth phases can be usefully identified using quantitative data on changes in firm size and employment, qualitative data uncover the endogenous processes and sector and location-specific sources of technical change that construct evolutionary (i.e. qualitative learning outcome based) change, which may otherwise remain buried and tends to get lost upon aggregation (Yin, 2009). A process study can allow for the exploration of descriptive and causal components and characteristics of phenomena which are difficult to separate and bound (Yin, 2009), such as for example firm learning activities, and the importance of different knowledge embodied in products, processes, firms and people for cluster dynamism. To investigate relationships between the rich concepts such as technological heterogeneity and global and local learning described in Section "Conceptual Framework: Cluster Evolution and Features of Creative Industries", we have posed "how" and "why" questions because of the wide variety of possible sources and factors underlying cluster evolution (such as firm learning activities, changes in products and systems, regional knowledge base augmentation, changes in technology areas, sectoral specificites and interdependencies) and the many different possible outcomes (for example, adaptation, transformation, learning, innovation and the emergence of specializations and paths). The research tries to take a longitudinal rather than cross-sectional perspective. A problem with using interviews to collect qualitative data on processes that occurred in the past is that people may not remember them accurately (Moser & Kalton, 1971). We tried to reduce this problem by also using data from secondary sources (Sinozic *et al.*, 2013).

A combination of theoretical and statistical sampling was used to select the firms. Mainly innovation-intensive firms were selected because the focus was on understanding qualitative change and to make analytic as opposed to statistical generalisations, i.e. technical change processes that interact with global and local changes and learning conditions. No firms that are mainly involved in re-sale were selected. It was difficult to fix the New Media sector because it changes rapidly and Nomenclature statistique des Activités économiques dans la Communauté Européenne (NACE) codes are not up to date with the changes in sectoral classifications. To remedy this, we stuck as close to previous studies as possible (in particular Lazzeretti *et al.*, 2008), and selected the following sectoral NACE categories: advertising (7311), film and video production (5911), ICTs (7311; 6209) and publishing (1812). Based on these selection criteria, in 2013, the New Media cluster in Vienna had a total of 480 firms, from which we interviewed 25 firms, four local industry experts and one central policy-maker. Firm interviews were carried out face-to-face with general managers, and lasted between 1 and 3 hours. The results were generalized to cluster features (not to the population), such as sources of technical change, learning processes and barriers and supports to the operationalization of technology as provided by the framework, using a mixed-methods approach combining qualitative analyses such as ordering and coding and basic descriptive statistics.

The unit of analysis is the cluster and its firms, defined as interconnected firms and institutions in a similar field which are also geographically concentrated (Porter, 1998) because it is broad enough to incorporate elements, processes and their interactions that are alluded to in the framework. We selected the New Media cluster, a subsector of creative industries, in Vienna because it is a dynamic cluster that has attracted research and policy attention (Ratzenböck *et al.*, 2004; ZEW, 2008). We designed a semi-structured interview guideline based on the concepts and conceptual relationships provided by the framework.

Results and Analysis

Cluster Technological Heterogeneity

This section highlights some of the main features of technological heterogeneity, innovation and sources of change in the New Media cluster in Vienna. New Media firms in Vienna are a heterogeneous group which has evolved from specializations in advertising, ICTs (such as software and gaming), design, publishing, communications, consulting and data management. The convergence of these areas into New Media activities (such as digital advertising, design and creation of application for digital devices, and, in some cases, their subsequent bundling with consulting and strategic services) showed its beginnings in the early 1990s (KMU Forschung *et al.*, 2010), about a decade later than the pioneer New Media firms in London (Lash, 2001) and thus at a later stage in the global industry life cycle. The systemic and interdependent nature of this sector with technical changes mediated via product specification routes continues to dominate firm operations, as one interviewee from a software design firm said: "We are continuously trying to improve upon multi-touch devices, tools, and modelling, which is bringing us closer to other sectors such as advertising and marketing".

Risks and opportunities accompanying technological heterogeneity in the cluster were confirmed by the stages of firm development (Table 1). The main points that stand out from these data are that the highest proportion of firms considered itself in the transformation stage (48%) and in the growth stage (40%). Firms referred to the transformation

Table 1. Stage of firm development

Stage of firm development	Per cent (freq)
Emergence stage	4 (1)
Growth stage	40 (10)
Maturity stage	8 (2)
Adaptation/transformation stage	48 (12)

Note: $N = 25$.

stage as when there is a change in organizational structure coupled with a change in product orientation. For example, one firm was reorienting its previous new media/advertising activities for the soft drinks industry, to the design and co-production of a novel product environmentally friendly car, in combination with an increase in the consulting component of its new media services. The growth stage was characterized as a stage following trans- formation, indeed the majority of the firms in the sample reported to have gone through various stages of restructuring followed by growth and restructuring to avoid decline.

Adaptation to technical change also took the form of making existing software and design processes compatible with novel digital devices and products (such as, for example, smartphones, tablets, touchscreens and cloud computing). For example, a gaming firm interviewed was previously developing games for consoles, and presently developing games for mobile devices. Our interviewee stated:

> (...) before you would go to a shop and take a game off the shelf and play it on your PC, you would finish the game and then go and buy a new one. Now people are playing games on their mobile phones, which they are getting for free, and they play it for years. We no longer make money by selling the game, but by selling in-game components.

Adaptation was also a process occurring as firms were modifying their existing software to different sectors such as changing markets from telephony to the healthcare sector. A further set of firms was renewing its services by changing their client base, based on changes in cul- tural interests: "A few years ago we were making strictly new media platforms, for new media communities, now we are changing to political interest communities."

A set of adjustment processes concerned the adaptation of internal firm structures to technological change. Changes in new media technologies not only affect products and services, but also the ways in which individuals in firms work and carry out their oper- ations and practices, sometimes radically enough to justify modifications of organizational structure. One of our interviewees stated:

> We transformed our firm to fit the changes in work processes. At first we were one large New Media firm, now we are several small highly specialised firms which are stand-alone entities based on client needs. It has become much more complicated to navigate the convergent tech- nologies that are currently on the market, and they are constantly changing, and it is easier for us to do that with small teams. So far our new structure has proven to be good, but new tech- nological changes may come which may mean we have to restructure again.

The supplier-dominated nature of technological heterogeneity in the cluster is further indi- cated in Table 2, which shows that the dominant innovation activities are the introduction or improvement on products or services (88%), and the use of a new or improved process, component or material (76%). One interviewee saw this as a constraint on creativity:

> A big part of our work is making our products fit the requirements of Facebook, Google, and Apple. But is that creative, is that innovative? Because these products are always changing, we do not know if any of our services will survive, or even if they are creative.

Lack of a stable relationship between technology and firm performance is also reflected in the vagueness of what constitutes a "New Media service" and how much a firm should charge its

Table 2. Innovation activities

Type of innovation	Per cent
Introduction or improvement on products or services	88
Introduction of a new product to market	60
Use of new or improved process, component or material	76
Use of new or improved strategy	64
Use of new or improved organizational structure	56
Introduction of a new or improved marketing concept	40

Note: $N = 25$.

customers for it. In this sector, service production overlaps with service creation because of tight process interdependencies with client needs which are continuously changing. These operational dynamics make it difficult to forge a local market that both firms and customers can depend upon: "(…) the customers do not know what they are getting when they ask for a new media service, why one company asks 500 Euros for a service, and another company asks 50,000 Euros for the same request." The prevalent insecurities in linking services to prices are a feature of the uncertainty that accompanies technological change more generally and the cluster specifically. The same interviewee added: "A local customer knows what they will get from a traditional advertising company in Vienna, but not from a new media firm."

Despite the disruptive nature of technological heterogeneity, we also found evidence of some stability in intra-sectoral networking and convergence upon related technological problems. Table 3 shows the self-reported network partners of the interviewed firms. The largest majority (45%) of partners are firms in the same sector. The main reasons for interactions with other firms given by our interviewees were know-how on changing (IT) system components for practices internal to the firm, and for individual New Media projects buying-in complementary design, content and technical skills. The relative importance of local networking (Table 4) also provides some evidence of the existence of the distributed nature of technological capabilities across same-sector firms, a favourable condition for sustained localized growth.

Repeated interactions over time have increased trust between cluster members creating conditions that provide some constancy in relationships. An important feature of interconnections between clustered firms is that over time as the cluster evolved some aspects of collaboration have become more virtual (digital). As one interviewee put it: "At first when

Table 3. Type of innovation-relevant network partners

Type of partner	Per cent (freq)
Subcontractor	4 (3)
Client/customer	17 (14)
Firm in same sector	45 (38)
Firm in different sector	2 (2)
University	20 (17)
Government agency	12 (10)

Note: $N = 25$. Total network partners in sample where type was specified: 84.

Table 4. Location of innovation-relevant network partners

Location	Per cent (freq)
Vienna region	49 (42)
Austria nationally	12 (10)
EU	15 (13)
Global	31 (27)

Note: $N = 25$. Total network partners in sample where type was specified: 86.

we did not know each other well we met face-to-face, until we got more experience with working together and we could carry out our work virtually, now we do a lot more over cloud computing." This finding resonates with the well-known view that through experience the tacit component of knowledge is embodied reducing the need for face-to-face interaction (Nonaka & Takeuchi, 1995).

Institutionalization (or fomalization) of interactions between firms has added certainty to the cluster and improved network stability. One of our interviewees described this in the following way:

> At the beginning everything was project dependent and we worked with each other informally, and we did not protect ourselves or make contracts. Now over time we are changing this, we are making our collaborations more formal, with contracts, and we work together in a more structured way. For example, we are making lists of all our partners, and in which steps of the process of service development they are important.

This echoes Winter's (1984) view of exploration and Bergman's (2008) view of early stages of cluster formation characterized by experimental relations due to cluster knowledge being too heterogeneous to formalize, and networks unstable, gradually stabilizing and becoming more formal characterizing later stages of cluster development.

Table 4 shows that almost half of all networking partners were located in Vienna and almost one third globally. Interaction with local partners was important because of the sufficiency of local capabilities, and the role of face-to-face contact in project-based work practices. As one interviewee put it: "Vienna, together with Berlin, is second in Europe for New Media services (London is first)."

Thirty-one per cent of innovation partners are global (Table 4), indicating a new media cluster with a global network (national- and European Union (EU)-level partners are less important), and a domination of firms outside the EU for this sector. Global firms were important sources of technical products (such as systems and programmes) and were mainly located in the USA. The following section presents some findings on the changing importance of local, national and international factors for learning and performance for the firms and the cluster.

Changing Importance of Regional, National and International Factors for Firm and Cluster Learning Over Time

In addition to the importance of products and processes for the existence and exploitation of technological heterogeneity in the cluster, change processes are shaped by factors such as skilled individuals, universities, investors, public funding agencies, networks and regulation at the regional, national and international levels. A tenet of cluster life cycle theory is that the characteristics of these factors affecting firm and cluster performance change as the cluster matures (Menzel & Fornahl, 2010). In evolutionary theories of technical change, these factors are important because they shape technological capabilities and local learning conditions within which capabilities are created and maintained (Bell & Albu, 1999).

Skills drawn from local and national levels are the principal factors underlying firm and cluster performance for the New Media sector (Table 5). Skills and experience of individuals have risen in importance because of increasing demand, and the unstable nature of technical change in this sector raising work process complexity. New media firms in Vienna place much importance on local graduates for recruitment of IT, designers and texters, implying that it is these types of know-how that are the ingredients for capabilities and the improvement of firm performance. The important roles of local culture, taste and aesthetic values in new media services were said to be difficult or impossible to obtain

Table 5. Importance of regional, national and international factors for firm development

	Degree of importance	Regional[a]		National[b]		International[c]	
		Previous 3–5 years (%)	In 2013 (%)	Previous 3–5 years (%)	In 2013 (%)	Previous 3–5 years (%)	In 2013 (%)
Skills	Very important	48	64	40	56	20	32
	Important	16	20	16	24	20	28
Universities/ research institutes	Very important	2	16	4	12	0	4
	Important	3	12	8	12	8	8
Demand	Very important	28	36	40	44	20	44
	Important	28	8	24	4	16	12
Other firms	Very important	24	32	20	28	20	24
	Important	36	32	36	32	12	36
Investors/ finance	Very important	4	8	4	4	0	4
	Important	20	8	16	8	4	16
Networks	Very important	32	44	28	36	12	28
	Important	16	24	24	24	4	24
Subsidies	Very important	12	20	12	12	8	12
	Important	12	20	12	20	0	8
Regulation	Very important	4	4	8	8	4	8
	Important	0	0	4	4	0	12
Directives	Very important	0	0	4	8	4	4
	Important	0	0	4	4	0	12

[a]$N = 25$ for all factors apart from "universities and research institutes" and "investors and finance" for which $N = 24$.

[b]$N = 25$, except for the following factors: "universities and research institutes" ($N = 22$), "other firms" ($N = 24$), "subsidies" ($N = 23$), "regulation" ($N = 22$) and "directives" ($N = 22$).

[c]$N = 25$, except for the following factors: "universities and research institutes" ($N = 22$), "other firms" ($N = 24$), "subsidies" ($N = 23$), "regulation" ($N = 22$) and "directives" ($N = 22$). All missing values are due to "no answer".

from international labour. The temporary type of employment which is required for project-based work in new media is more easily achieved under conditions of physical proximity. As one interviewee put it:

> I employ more or less the same people for the last ten years, the same designers, texters and programmers, but when there is no project my firm has only one employee (only me). I can only do that because all of these people live in Vienna and they can jump on a project at short notice.

The cluster is also characterized by the importance placed upon local demand and information sourced via networks (Table 5). Over time, the local orientation in terms of these factors has increased, indicating a rising importance of the local cluster for the firms over time. Despite the decline in communication costs via digital technologies, the importance of culture, language and practice in community-based processes involved in creative work such as digital advertising and marketing strategies, as well as the maintenance of personal networks that inter-organizational projects depend upon, are key features of how capabilities in this cluster are created and maintained and the processes through which uncertainty is reduced.

Table 5 also presents the change in importance of national factors for the firms. Of these, several important findings stand out. First, the same national factors—skills, demand, and other firms and networks—shaped firm activities as regional factors indicating that the operative region for New Media firms in Vienna is all of Austria. The increase in importance of networks at the national level (a total of 52% of firms found national networks as very important or important previously, and 60% presently) indicate a formation of a national network over time.

International-level factors are not perceived as important for new media firm activities as factors at the regional and national levels. Of the international factors, the most importance was given to skills, international demand, other firms globally and international firm networks. These three sets of factors are echoed as most important at all three (regional, national and international) levels, confirming that they are most connected to firm competitiveness overall. These findings confirm the capabilities and network perspectives put forward by, amongst others, Ter Wal and Boschma (2011), who argue that the interplay between knowledge, the position of the firm in the regional network and its industry underlie the evolution of clusters.

International networks which are particularly important are those associated with IT, advertising and design. International networks are considered as an important route for keeping up to date with fast-changing products and services, as are relationships with other global firms. International IT skills are gaining in importance because of the shortage of programmers in Vienna and in Austria nationally. The international competition for IT skills has made recruitment more difficult and driven wage prices up (according to the firms interviewed).

Table 6 further confirms the importance of Vienna and Austria as dominant levels for New Media firms. The low level of entry into global markets is indicative of, on the one hand, the importance of local territory for the creation and delivery of services, and on the other hand, difficulties with internationalization and expansion to the global stage because of a variety of barriers, as one interviewee from a digital design firm stated:

Table 6. Market areas

Geographical level	Average of market areas in per cent
Regional (Vienna)	32.96
National (Austria)	40.05
EU	22.93
Global	3.22

Note: $N = 25$.

> I need to open up an office in China now, to expand to that market, but I don't even know where to begin. We don't speak the language, and all my employees want to remain in Vienna, no one wants to relocate to China.

The EU level, specifically the countries Germany and Switzerland, are the third-most relevant markets for New Media firms, following Vienna and Austria. The low cultural and language barriers in Switzerland and Germany for Austrian firms make it easier to expand to those markets than to other countries in the EU.

In order to explore changes in spatial boundaries of the cluster, we further asked firms about their views on the importance of regional, national and international factors for the "cluster", to understand whether certain factors indirectly supported or constrained the population of firms and institutions in New Media in Vienna. Skills supply, in particular IT and design, were the considered the most important cluster resources. Their importance has grown over time because of the problem-solving required to deal with the complexity of technologies and services, the increase in specialization (IT skills and design) and the fast-paced changes in the new media sector overall which call for flexible and highly skilled individuals who are able to keep up with the changing demands of the marketplace. High importance was placed on regional networks and firms, for the same reasons as skills (they are important sources of know-how).

Regional subsidies and universities and research institutes were considered as important for the cluster (but not for the firms directly). The regional science base and subsidies act as important framework conditions for sustaining the sector, and are understood as having a very important but indirect role for New Media in Vienna. Moreover, these factors are also considered to have not changed in importance over time, but as relatively stable and unchanging in their importance. Finance was repeatedly stated as more important for the sector than for firms individually, but because of its limited availability (some parts of the gap being filled by government subsidies), their importance for the cluster is perhaps overstated.

Both international regulation and international directives have grown in importance for the cluster quite dramatically. This was explained by the regulation of data protection, and intellectual property rights (IPRs) for IT products, most associated with US firms and their patenting activities. Patenting of intellectual property in this sector was described by two interviewees as a constraint to innovation because of the threat of being sued, and the shortage of insight into the US patent system by Viennese firms.

Conclusion

This paper has investigated technological heterogeneity in the New Media cluster in Vienna, and the local and global factors that have affected firm and cluster performance over time. In doing so, it focused on opening the "black box" of technological heterogeneity and its shaping over time, to uncover the finer details of and interdependencies within the diverse embodiments of technical change, and their influential, emergent, complex, systemic and partially unpredictable unfolding in the process of cluster evolution. In growing clusters whose performance is tightly interwoven with emergent technology areas such as digital design, a central feature of local technological heterogeneity are technological capabilities that support firms in the translation of different sources of technical change (such as systems and devices) into procedures and operations that sustain or enhance their performance. The adaptation of local learning conditions, such as the support of skills formation through education, is important for the lowering of technological uncertainty that heterogeneity brings about.

The relationship between technological heterogeneity and cluster evolution is not straightforward. Cluster performance is enhanced by technological heterogeneity if technological capabilities for its exploitation and local conditions for lowering the uncertainty of its unpredictable outcomes exist. The sources and elements of technological heterogeneity are diverse and continually changing. In supplier-dominated sectors such as New Media, sources of technical change travel via products, systems, platforms and people. These sources depend upon local technological capabilities for their operationalization and transformation. Even if technological capabilities to manage technological changes exist at certain times, the potentially long and error-ridden periods available to detect and manage learning means that their effect on cluster growth may be delayed. In other words, the exploitation of technological heterogeneity may not directly overlap with a cluster growth phase, but more likely it will delay it. If products and systems are rapidly changing, then firms focus on incremental improvements and adjustments. Firms may also be forced to change strategies, such as adapting organizational structures (e.g. from departments to small teams) or applying the components of existing products and systems to different markets (e.g. from telephony to the healthcare sector).

Certain features of technological heterogeneity add stability to processes underpinning cluster growth. In network-dependent project-based work such as in New Media, the convergence of cluster firms on technological focal points is supported by the existence and stability of firm networks and regional demand within the same sector. Networks between firms possessing complementary resources such as for example programming, texting and digital design know-how are more important for enhancing performance than diverse but unrelated resources.

Technological heterogeneity within the cluster and outside the cluster are not easily separated, rather, technologies connect clusters to other sectors and geographies. In supplier-dominated sectors such as New Media, diversity is highly interdependent with radical and partially unpredictable innovation external to the cluster (e.g. changes in product platforms by US firms). The connections to other sectors, overlaps and movements towards the creation of novel technological focal points can increase instability for clustered firms by creating inconsistencies and difficulties affecting a variety of procedures and operations making the outcome for cluster growth phases unpredictable in the short run.

In the New Media cluster in Vienna the importance of resources such as skills, and information and resources obtained through networks, as well as local demand, has greatly increased over time. Of particular importance were skills to deal with the increasing complexity of fast-paced changing technologies and client needs. Insufficient supply of IT and programming skills at the local and national levels was an important barrier to innovation and growth. International skills and interactions with global firms (via products, processes and collaborative projects), international demand and networks have increased in importance for the cluster over time, intensifying its global interdependencies and broadening the supply of technologies. Those firms that had accumulated the relevant know-how and resources could capitalize more directly on these changes and support cluster growth. Over time, international regulation and directives have grown in importance for the sector, especially in the regulation of data protection and IPRs in the global IT industry (most associated with US firms). Learning conditions in the cluster were changing by global technical change processes. Less importance was given to creativity, and more importance was given to managing the changing technological demands from outside the cluster (in the form of changing products, platforms and processes) making work processes inside the cluster more complex and firm survival more difficult. Future research should look into how local conditions can support the accumulation of technological capabilities in New Media and help reduce technological uncertainty in cluster evolution.

Acknowledgements

We would like to thank three anonymous referees for their valuable comments on an earlier version of this article. We gratefully acknowledge the support of our project partners from the University of Kiel, Germany, the University of Hamburg, Germany, University of Bremen, Germany, Lund University, Sweden, University of Agder, Norway, Vienna University of Economics and Business, Austria, Charles University in Prague, Czech Republic, Silesian University in Opava, Czech Republic, University of Ostrava, Czech Republic, University of Neuchatel, Switzerland, and the INSEAD Policy Initiative, Abu Dhabi.

Funding

This work was supported by the European Science Foundation European "Cluster Life Cycles Project" and by the Austrian Science Fund (FWF) [Grant number I 582-G11], and coordinated by Professor Robert Hassink, University of Kiel.

Note

1. For a more detailed review, see Sinozic *et al.* (2013).

References

Audretsch, D. B. & Feldman, M. (1996) R&D spillovers and the geography of innovation and production, *American Economic Review*, 86(630), pp. 630–640.
Barrett, F. J. (1998) Creativity and improvisation in Jazz and organizations: Implications for organizational learning, *Organization Science*, 9(5), pp. 605–622.
Bell, M. & Albu, M. (1999) Knowledge systems and technological dynamism in industrial clusters in developing countries. *World Development*, 27, pp. 1715–1734.
Bell, M. & Pavitt, K. (1993) The development of technological capabilities, Chapter 4, in: I. ul Haque (Ed) *Trade. Technology and International Competitiveness*, pp. 69–101 (Washington, DC: EDI Development Studies, The World Bank).

Bell, M. & Pavitt, K. (1997) Technological accumulation and industrial growth: contrasts between developed and developing countries, Chapter 4, in: D. Archibugi and J. Michie (Eds.) *Technology, Globalisation and Economic Performance*, pp. 83–137 (Cambridge: Cambridge University Press).

Bergman, E. M. (2008) Cluster life-cycles: an emerging synthesis, in: C. Karlsson (Ed) *Handbook of Research in Cluster Theory*, pp. 114–132 (Cheltenham: Edward Elgar).

Boschma, R. A. & Frenken, K. (2006) Why is economic geography not an evolutionary science? Towards an evolutionary economic geography, *Journal of Economic Geography*, 6, pp. 273–302.

Boschma, R. A. & Frenken, K. (2009) Some notes on institutions in evolutionary economic geography, *Economic Geography*, 85(2), pp. 151–158.

Boschma, R. A., & Lambooy, J. G. (1999) Evolutionary economics and economic geography, *Journal of Evolutionary Economics*, 9, pp. 411–429.

Braunerhjelm, P. & Feldman, M. (2006) *Cluster Genesis: Technology-Based Industrial Development* (Oxford: Oxford University Press).

Cooke, P. & Lazzaretti, L. (Eds) (2008) *Creative Cities, Cultural Clusters and Local Economic Development* (Cheltenham: Edward Elgar).

DCMS. (1998) *Creative Industries Mapping Document* (London: Routledge).

DeFillippi, R. J. & Arthur, M. B. (1998) Paradox in project-based enterprise: The case of film making, *California Management Review*, 40(2), pp. 125–139.

De Propris, L. & Lazzeretti, L. (2009) Measuring the decline of a Marshallian industrial district: The Birmingham jewelry quarter, *Regional Studies*, 43, pp. 1135–1154.

De Propris, L. & Wei, P. (2007) Governance and competitiveness in the Birmingham jewellery district, *Urban Studies*, 44, pp. 2465–2486.

Dosi, G. (1982) Technological paradigms and technological trajectories: A suggested interpretation of the determinants of technical change, *Research Policy*, 11(3), pp. 147–162.

Dosi, G. (1988) The Nature of the Innovation Process, in G. Dosi, C. Freeman, R. R. Nelson, G. Silverberg and L. Soete: *Technical Change and Economic Theory*, pp. 221–238 (London: Pinter).

Florida, R. (2002) *The Rise of the Creative Class* (New York: Basic Books).

Fornahl, D., Henn, S., & Menzel, M.-P. (2010) *Emerging Clusters: Theoretical, Empirical and Political Perspectives on the Initial Stage of Cluster Evolution* (Cheltenham: Edward Elgar).

Freeman, C. (1982) *The Economics of Industrial Innovation*, 2nd edn (London: Pinter).

Frenken, K. & Boschma, R. A. (2007) A theoretical framework for evolutionary economic geography: industrial dynamics and urban growth as a branching process, *Journal of Economic Geography*, 7, pp. 635–649.

Grabher, G. (2001) Ecologies of creativity: the village, the group, and the heterarchic organisation of the British advertising industry, *Environment and Planning A*, 33, pp. 351–374.

Hassink, R. & Shin, D.-H. (2005) The restructuring of old industrial areas in Europe and Asia, *Environment and Planning A*, 37, pp. 571–580.

Hatch, M. J. (1999) Exploring the empty spaces of organizing: How improvisational Jazz helps redescribe organizational structure, *Organization Studies*, 20(1), pp. 75–100.

Hobday, M. (1995) East Asian latecomer firms: Learning the technology of electronics, *World Development*, 23(7), pp. 1171–1193.

KMU Forschung Austria., Zentrum fuer Europäische Wirtschaftsforschung (ZEW), and Joanneum Research. (2010) *Vierter Österreichischer Kreativwirtschaftsbericht* (Vienna: creativ wirtschaft Austria).

Lash, S. (2001) New media: Technological creativity or corporate consultants? International Workshop on the Socio-Economics of Space, University of Bonn, Bonn.

Lazzeretti, L. (2008) The cultural districtualisation model, in: P. Cooke & L. Lazzaretti (Eds) *Creative Cities, Cultural Clusters and Local Economic Development*, pp. 93–121 (Cheltenham: Edward Elgar).

Lazzeretti, L. (2012) *Creative Industries and Innovation in Europe: Concepts, Measures and Comparative Case Studies* (Oxon: Routledge).

Lazzeretti, L., Boix, R. & Capone, F. (2008) Do creative industries cluster? Mapping creative local production systems in Italy and Spain, *Industry and Innovation*, 15(5), pp. 549–567.

Lorenzen, M. & Frederiksen, L. (2008) Why do cultural industries cluster? Localization, urbanization, products and projects. In P. Cooke & L. Lazzaretti (Eds) *Creative Cities, Cultural Clusters and Local Economic Development*, pp. 155–179 (Cheltenham: Edward Elgar).

Martin, R. & Sunley, P. (2006) Path dependence and regional economic evolution, *Journal of Economic Geography*, 64(4), pp. 395–437.

Maskell, P. & Kebir, L. (2005) *What qualifies as a cluster theory?* DRUID Working Paper, Department of Industrial Economics and Strategy, Copenhagen Business School, Copenhagen.

Maskell, P., & Malmberg, A. (2007) Myopia, knowledge development and cluster evolution, *Journal of Economic Geography*, 7, pp. 603–618.

Menzel, M.-P. & Fornahl, D. (2010) Cluster life cycles—Dimensions and rationales of cluster evolution, *Industrial and Corporate Change*, 19(1), pp. 205–238.

Moser, C. A. & Kalton, G. (1971) *Survey Methods in Social Investigation* (Surrey: Ashgate).

Nelson, R. R. & Winter, S. G. (1982) *An Evolutionary Theory of Economic Change* (Cambridge, MA: Harvard College).

Nonaka, I. & Takeuchi, H. (1995) *The Knowledge-Creating Company: How Japanese Companies Create the Dynamics of Innovation* (New York: Oxford University Press).

Pavitt, K. (1984) Sectoral patterns of technical change: Towards a taxonomy and a theory, *Research Policy*, 13, pp. 343–373.

Porter, M. E. (1998) Clusters and the new economics of competition, *Harvard Business Review*, 76, pp. 77–90.

Power, D. & Scott, A. (Eds) (2004) *Cultural Industries and the Production of Culture* (London: Routledge).

Pratt, A. C. (1997) The cultural industries production system: A case study of employment change in Britain, 1984–91, *Environment and Planning A*, 29(11), pp. 1953–1974.

Pratt, A. C. (2008) Creative cities: The cultural industries and the creative class, *Geografiska Annaler: Series B, Human Geography*, 90(2), pp. 107–117.

Ratzenböck, V., Demel, V., Harauer, R., Landsteiner, G., Falk, R., Leo, H., & Schwarz, G. (2004) *Endbericht: Untersuchung des Ökonomischen Potenzials der "Creative Industries" in Wien* (Stadt: Wien).

Resch, A. (2008) Anmerkungen zur langfristigen Entwicklung der "Creative Industries" in Wien, Chapter 1 in Mayerhofer, in: P. Peltz & A. Resch (Eds) *"Creative Industries" in Wien: Dynamik, Arbeitsplätze, Akteure*, pp. 9–33 (Vienna: LIT Verlag).

Rosenberg, N. (1982) *Inside the Black Box: Technology and Economics* (Cambridge: Cambridge University Press).

Scott, A. J. (1997) The cultural economy of cities, *International Journal of Urban and Regional Research*, 21, pp. 323–339.

Simmie, J. (2012) Path dependence and new technological path creation in the Danish wind power industry, *European Planning Studies*, 20(5), pp. 753–772.

Sinozic, T., Auer, A. & Tödtling, F. (2013) 'Growth and transformation in Vienna's creative industries', Paper presented at the Regional Studies Association European Conference 2013: "Shape and be Shaped: The Future Dynamics of Regional Development", University of Tampere, Finland, May 5–8, 2013.

Storper, M. (1997) *The Regional World: Territorial Development in a Global Economy* (New York: Guilford Press).

Swann, G. M. P. (2002) Towards a model of clustering in high-technology industries, in: G. M. P. Swann, M. Prevezer & D. Stout (Eds) *The Dynamics of Industrial Clustering*, pp. 52–76 (Oxford: Oxford University Press).

Sydow, J. & Staber, U. (2002) The institutional embeddedness of project networks: The case of content production in German television, *Regional Studies*, 36(3), pp. 215–227.

Ter Wal, A. L. J. & Boschma, R. A. (2011) Co-evolution of firms, industries and networks in space, *Regional Studies*, 45(7), pp. 919–933.

Trippl, M., Tödtling, F., & Schuldner, R. (2012) The geography of creative and cultural industries in Austria, Chapter 4, in: L. Lazzeretti (Ed) *Creative Industries and Innovation in Europe: Concepts, Measures and Comparative Case Studies*, pp. 86–102 (Oxon: Routledge).

Winter, S. G. (1984) Schumpeterian competition in alternative technological regimes, *Journal of Economic Behaviour and Organisation*, 5, pp. 287–320.

Yin, R. K. (2009) *Case Study Research: Design and Methods* (Thousand Oaks: Sage).

ZEW (Zentrum für Europäische Wirtschaftsforschung) & Fraunhofer Institut System-und Innovationsforschung (ISI). (2008) *Beitrag der Creative Industries zum Innovationssystem am Beispiel Österreichs*, (Fraunhofer ISI: Karlsruhe and ZEW: Mannheim).

Creative Cluster Evolution: The Case of the Film and TV Industries in Seoul, South Korea

SU-HYUN BERG

Department of Geography, University of Kiel, Kiel, Germany

ABSTRACT   *Can the concept of co-evolution help to analyse and explain the dynamics of creative industries? This article tackles the question by investigating the film and TV cluster in Seoul, South Korea. The analysis of the 35 semi-structured interviews confirms the dynamics of the film and TV industries in Korea. First, Hallyu began with the export of Korean TV drama series across East Asia. The state deregulation and neo-liberal reforms during the 1990s in Korea boosted an explosion of the export of the Korean film and TV industry. Second, the core of the film and TV production is concentrated within Seoul, while dispersion of those industries occurred in Gyeong-gi province. Third, from an institutional perspective, tensions between the central government and the film and TV industry can be observed, which have been intensifying since 2006. This paper concludes that particularly co-evolution could potentially be an important concept to explain and analyse dynamics in creative industries.*

1. Introduction

Recently, the popularity of South Korean cultural goods—movies, TV programmes, pop music and computer games—has rapidly increased. This phenomenon is referred to as the Hallyu (translated into English as the Korean Wave) phenomenon (Cho, 2012; Jang & Paik, 2012). For instance, Korean rapper PSY's "Gangnam Style" became the first music video clip in the history of YouTube to reach one billion views, on 22 December 2012. "Gangnam Style" was not only the first single video to record one billion views since the launch of YouTube in 2005, but it also achieved this record in just 161 days (Koehler, 2013). U.N. Secretary General Ban Ki-moon met PSY at the United Nations Headquarters and he made a comment that PSY has an "unlimited global reach". (Jin, 2012, p. 5) Furthermore, Korean director Kim Ki-duk's film "Pieta" won the Golden Lion, the award for the best film, at the 69th Venice Film Festival on 10 September 2012. Kim's Golden Lion marked the latest in a series of Hallyu phenomena on the international stage. Recently, President Park Geun-hye—the first female president of the Republic of Korea—and her new administration pledged to build a new creative economy on the basis of the booming, globalizing creative industries in Korea. Park aims to bring about an economic revival by fostering a creative economy. Many studies have been done on the economic factors of Hallyu –such as Hallyu impacts on film and TV programme exports (Ryu, 2013), tourism growth (Jang & Paik, 2012) and technology innovation (Cho, 2005). However, only few studies have been done on the spatial evolution of the leading creative industries—for instance, the film and TV production industries—which are benefiting from the Hallyu phenomenon.

Evolutionary Economic Geography (EEG) typifies and explains the development of industries through time (Boschma & Frenken, 2006; Boschma & Martin, 2010). EEG potentially has some key explanatory notions—such as windows of locational opportunity and path creation, path dependence and lock-ins, related variety and branching and co-evolution. In particular, co-evolution can potentially be an important notion for explaining and analysing the dynamics of creative industries, as it deals with the relationship and interaction between an industry and its institutional settings.

Therefore, the primary goal of this paper is to analyse whether notions of EEG can help to analyse and explain creative cluster evolution by investigating the film and TV industries in Seoul, South Korea.

This paper is based on extensive fieldwork from February to April 2013. Semi-structured interviews, discussion with experts, field observation and participation in forums, seminars and conferences were conducted during a research trip to South Korea. I stayed for seven weeks in order to gain practical experiences in the film and TV industries. In this stage, 35 semi-structured interviews[1] were made and recorded. Nineteen interviews were carried out with representatives of institutional sectors –including public research institutes, the Ministry of Culture, Sports and Tourism, the Korean Film Council (KOFIC) and the Korea Creative Content Agency (KOCCA)—and 16 interviews with firms—such as Korean Broadcasting System (KBS), MBC, CJ entertainment and Lotte Cinema. On average, the interviews took about 60 minutes. The interview partners included film directors, producers, technicians, policy-makers and industry experts.

With the help of the interviews, this paper first examines the relation between the Hallyu phenomenon and the film and TV industries. Second, the paper discusses the locational patterns of the film and TV industries in Korea as leading industries of the Hallyu phenomenon. Finally, the paper critically evaluates the co-evolution of the film and TV industries and their institutional environment.

The paper proceeds as follows. It begins with co-evolution by focusing on the conceptual and theoretical underpinnings. This theoretical proposition is then empirically explored in the following sections based on the film and TV industries in South Korea. Findings show that (1) Hallyu had a significant impact on the exports of the film and TV industries; (2) the film and TV industries in Korea shifted dispersion outside of Seoul; (3) co-evolution reveals matches and mismatches over time. The concluding section sums up the main-related theoretical and empirical implications.

2. Co-evolution

The concept of co-evolution derives from evolutionary thinking; it can be applied to understand the dynamics of economies in space and time. Recently it has been discussed by economic geographers (Boschma & Frenken, 2006; Martin & Sunley, 2006; Boschma & Martin, 2010; Ter Wal & Boschma, 2011).

The term "co-evolution" was first used in biology and refers to a condition in which "two evolving populations co-evolve if and only if they both have a significant causal impact on each other's ability to persist" (Murmann, 2003, p. 210). However, many economic geographers felt a need for a restricted understanding of the concept of co-evolution, in order to move from the statement that everything is co-evolving with everything else to the identification of what is co-evolving with what, how intense this process is and whether indeed there is a bidirectional causality (Malerba, 2006, pp. 17–18; Ter Wal & Boschma, 2011, p. 920). Martin and Sunley (2006, p. 430) pointed out that "forms of co-evolution in which there are mutually constitutive interactions and feedbacks between firms and other institutions are to some degree place-specific, and that these interactions occur simultaneously across several different scales".

In a co-evolutionary perspective, it is not only firms and industries, but also in a broader sense the institutional setting of firms and industries, that can affect the dynamics of regional economies (Nelson, 1994; Murmann, 2003). The recognition of the importance of institutions in the study of EEG is not completely new (Boschma & Frenken, 2009). One classic example of the importance of co-evolution is Murmann's (2003) study on the dye industry in the nineteenth century, when Germany managed to carry out essential institutional changes (in education and regulations) to support the successful emergence of this industry, whereas the UK and the US did not.

Institutions can be defined as "formal regulations, legislation, and economic systems as well as informal societal norms that regulate the behavior of economic actors: firms, managers, investors, workers" (Gertler, 2004, p. 7). The objective of the study of institutions in an EEG framework is to throw light on the dynamic interplay between institutional change and industrial dynamics in economic development. In addition, institutions are not pregiven and fixed, but co-evolving with technologies and markets (Malmberg & Maskell, 2010, p. 396). Institutional settings thus can be acknowledged as regional, national and supranational levels of markets, regulations and infrastructures. Schamp (2010, p. 446) differentiated between three institutional forms, namely; (1) the firm (or the regulatory systems prevalent in the firm that Nelson and Winter, 1982, call routines), (2) the "social" regulatory systems affecting firms (which often exist at the national or even international level) and (3) the region as a specifically institutional construct that is a major focus of geographical interest. The focus of my empirical research is the case of the Korean film and TV industry, in which the public institutions played a substantive role by deploying development-oriented policies over the last two decades. This article hence attaches more importance to public institutions, less to private institutions.

Certain economic geographers tried to employ the notion of co-evolution to describe and explain various dynamics of spatial economic development in manufacturing industries (Menzel & Fornahl, 2010; Iammarino & McCann, 2010; Schamp, 2010; Boschma & Frenken, 2011). Recent studies, however, go one step further. They show that using the co-evolution concept helps to explain how changes in the institutional environment and the development of creative industries are linked to each other and mutually influence each other through time (Rantisi, 2004; Banks & Potts, 2010; Berg & Hassink, 2014).

The film and TV industry is categorized into production chains with five stages: development, shooting, post-production, distribution and exhibition (Dahlström & Hermelin, 2007). The production of films and television dramas is an especially prodigious and complex project that requires major funding and various actors for value creation. This project-based, flexibly specialized organization of value creation activities of the film and TV industry leads the industry to a geographical agglomeration in clusters like Hollywood and Bollywood (Lorenzen & Mudambi, 2012). Consumption of film and TV programmes takes place in an international market. Furthermore, film production has become more mobile as a result of technical developments and reduced costs in travel and communication. Given the nature of the industry, film and TV production is integrated into an international and globalized institutional environment (Miller et al., 2005).

From this perspective, the concept of co-evolution can potentially help to analyse and explain the dynamics of the film and TV industries in Korea, to which this paper will now turn.

3. Empirical Studies of the Film and TV Industries in Korea

The popularity of Korean TV dramas in China ("What is Love All About?" in 1997) and Japan ("Winter Sonata" in 2003) ignited the Hallyu phenomenon of the increasing popularity of Korean music, film and video games, etc. However, Hallyu would not have existed without the liberalization and development of the media in the 1990s. The structural and institutional changes of the 1990s in Korea, i.e. the state deregulation and neo-liberal reforms initiated the opening of its domestic market to the outside and encouraged Chaebols (large conglomerates such as Samsung, CJ Entertainment and Lotte Cinema) to enter the film and TV industry. Since then, the Korean film and TV industry has targeted domestic audiences and foreign consumers and has started to produce high-quality products (Ryoo, 2009).

The exports of Korean TV drama series and films doubled from 1999 to 2011. In addition, the number of film production companies increased from 2300 in 2002 to 4900 in 2011 (Kim E.S., 2012). The total export revenue of the Hallyu product recorded 553% growth between 2001 and 2011. The Hallyu exports reached USD 4.3 billion in 2011, a 34.9% increase over the previous year. Film exports rose by 16.5% and the export of TV programmes by 20.4%, compared to the year 2010. In addition, the number of employees increased—in the TV industry by 10.6% and in the film industry by 3.9% (Ministry of Culture, Sport and Tourism, 2013). Hallyu, at least in terms of the export and scale of the Korean film and broadcasting industry, has been in progress since the beginning of 2000.

Recently, Korean President Park Geun-hye put forward a vision of the "creative economy" as the core national agenda to revitalize the economy, presenting creativity and innovation as the key driving forces for the future growth of Korea.

> The concept of creative economy is not new, but the policy model of Korea is definitely the first of its kind in the world. Unlike the UK or Japan, Korea is approaching the creative economy as a paradigm shift of the whole nation, by converging creativity and ICT (Information and Communications Technology). It is a great chance for creative industries in Korea. The strong will of the central government motivates many stakeholders like me. (Film Investor A)

3.1. *Hallyu as a Critical Event?*

Hallyu is the term first mentioned by the Chinese media in the late 1990s in order to describe the rapid growth of Korean entertainment products (Hogarth, 2013). Opinions vary as to the exact origin of the Hallyu;

One Chinese newspaper at the first used the word "Hallyu" in 1999. At that time, Hallyu referred to the new cultural influence from South Korea. (KBS Hallyu expert H)
Even before the word "Hallyu" first appeared in 1999, Korean TV dramas were well known in China, Taiwan and Japan. In 2002, the FIFA World Cup took place in Korea and in Japan. Korean culture became very popular worldwide upon this opportunity. (Research fellow at the Korea Creative Content Agency I)
The 2002 FIFA World Cup Japan–Korea was the tipping point for the Hallyu phenomenon. The year 2002 is the point at which, after a series of small successes in South East Asia, Hallyu reaches a level where it begins to change dramatically and starts to have an impact on the exports of Korean cultural goods. (Professor at the department of Media Communication J)
The export of Korean films and TV programs increased dramatically after the FIFA world cup 2002 Japan–Korea. (government official at the Ministry of Culture, Sports and Tourism K)
Korean film gained much more attention benefiting from the Hallyu phenomenon, which started after the 2002 FIFA world cup. (Professor at the Department of Theatre L)

Despite the different views on the origin of Hallyu, most interviewees noted that the recent success of Hallyu has had a significant impact on Korean film and TV programme exports.

3.1.1. *Hallyu impact on the TV programme.* Korea's broadcasting industries have been exporting to Asian markets since the late 1990s and beyond Asia since the 2000s (Figure 1). Korean television programme exports increased by a factor of 40 between 1995 (USD 5.5 million) and 2011 (USD 220.9 million). Exports of TV programmes have been enjoying an annual growth rate of 9.8% since 2005. Korean drama series (87.8%) accounted for the largest share of the exports, above entertainment and documentaries. The majority of these Korean drama series were consumed in South East Asia, especially Japan (60.4%), Taiwan (12.5%) and China including Hong Kong (10.2%) (Ministry of Culture, Sport and Tourism, 2013).

3.1.2. *Hallyu impact on Korean films.* Along with the Hallyu phenomenon, there has been a significant increase in awards for Korean films, directors and actors. Korean cinema gained global recognition when Park Chan-Wook's Oldboy won the Grand Prix at the 2004 Cannes Film Festival.
Korean film director Kim Ki-duk's film Pieta won the Golden Lion, the award for best film, at the 69th Venice Film Festival (Koehler, 2013). In addition, the export revenue of Korean films increased by a factor of almost 65 between 1995 and 2005 (Shin & Stringer, 2005, KOFIC, 2003, 2005, 2008, 2012). The total export of Korean films reached USD 37 million in 2012, an 8.4% increase over the previous year (Figure 2).
The total export of Korean films reached USD 37 million in 2012, an 8.4% increase over the previous year. Furthermore, Korean films have maintained a market share of above 50% in the domestic market during the 2000s, representing their competitiveness against Hollywood films (Kwon & Kim, 2013). For instance, the market share of Korean films was 59.6% and that of foreign films was 40.4% in 2012 (Figure 3). The number of film production companies also increased from 2300 in 2002 to 5099 in 2012.
This dramatic rise of Korean films came from a deliberate move towards big-budget, high-risk/high-reward business. Hallyu has made many Korean filmmakers famous around Asia and the world, helping to increase cultural exports. During the 2000s, the Korean film industry experienced a wider series of changes (See Figure 2). Several factors have evolved in the past decade, including the reduction of the screen quota in

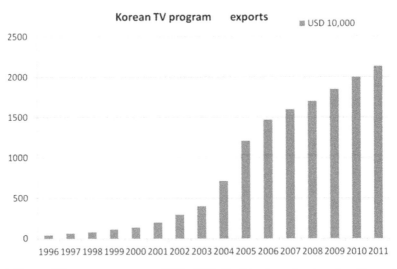

Figure 1. Korean TV programme exports 1996–2011. Draft by the author based on Korean Creative Contents Agency (2001, 2002, 2004, 2005, 2013).

2006 and the emergence of digital technological advances, education of personnel, redesigning of the related infrastructure, market-driven movies, creative production and the promotion of independent films.

There is as yet no generally accepted time point when exactly the Hallyu phenomenon started. However, there was a significant increase in exports of Korean film and TV productions after the FIFA World Cup Japan–Korea 2002. The FIFA World Cup Japan–Korea 2002 is not only the reason that Korean popular culture gained international recognition. The growth of social network systems, government export supporting policy and hybridization of domestic culture also played important roles. On the one hand, Hallyu

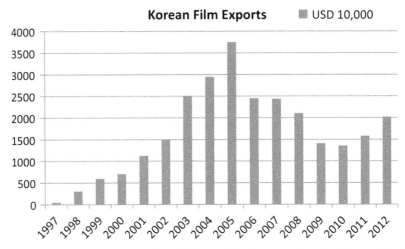

Figure 2. Korean film exports (year, USD10,000). Draft by the author based on Korean Film Council (2003, 2005, 2008) and Korean Creative Contents Agency (2013).

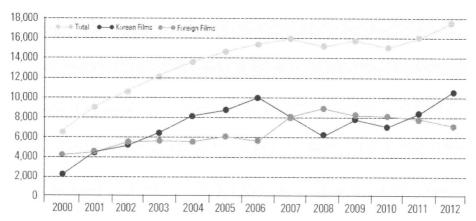

Figure 3. Total admissions of Korean and foreign films by year (in 10 thousands).
Source: Korean Film Council (2012) pp. 30.

is an example to illustrate the complexity involved in the cultural hybridization that works towards sustaining local identities within a global context. On the other hand, Hallyu is a result of the liberalizations of institutional structure and development of the media in the 1990s.

In sum, Hallyu focused attention on Korean cultural products—such as TV drama series, domestic film, video games, popular songs and food. In addition, Hallyu was a definitive tipping point for Korean creative industries, as they benefit from it. As Hallyu started in the late 1990s, the export of Korean film and TV products increased dramatically. Hallyu not only stimulated the growth of the film and TV industries in Seoul, but also the expansion outside of Seoul. Many film and TV productions are concentrated in the central business districts (CBD) in Seoul. However, small- and medium-sized productions have recently been moving out of Seoul and heading to the northern part of Gyeong-gi Province.

3.2. *From Concentration to Dispersion*

Seoul has three CBD in which office buildings are heavily concentrated, namely JBD (Jongno Business District in the city centre), GBD (Gangnam Business District) and YBD (Yeouido Business District). Around 70% of the office buildings in Seoul are situated in these three CBDs. (Figure 4).

JBD covers an area from Gwanghwamun and Seoul City Hall to Namdaemun and Seoul Station. It consists of the headquarters of major domestic companies, news agencies, law firms and financial institutions. In addition, the central government complex and a couple of government agencies are located in this area.

Until the 1970s GBD was a rice field. However, now this area is well known as the centre of education, business and shopping in Seoul. It encompasses office buildings from Gangnam-daero to Teheran-ro with its traffic flow of more than 200,000 people every day. In GBD, major companies in IT, manufacturing and pharmaceutical fields are located. Especially, luxury goods and fashion-related companies favour GBD for its name value. Most of the film production companies are located in the Gangnam area. In the 2000s, the Korean film industry experienced rapid growth with strong performances both in the domestic and foreign markets. The market share of local films in the Korean market continued to reach more than 50% in 2012 (Korean Film Council, 2013). The number of production companies increased from 715 in 2000 to 2465 in 2010. Over the same time period, the number of distributors and screens also increased from 259 to 575 and from 466 to 817. In addition, the major Korean enterprises with nationwide multiplex chains, represented by CJ and Lotte, have been striving to strengthen their market leadership in terms of production and distribution, as well as exhibition, since 2004. There are a total of 701 film production companies in Seoul and 53.3% of productions are located in GDB (Kwon, 2011, p. 116, Hwang, 2009).

Gangnam has a name value such as Hollywood in the USA. As a film production company, we have to locate here. (Film agency manager B)

Gangnam offers many production companies two attractions. One is the geographical proximity to the financial center YBD. It is especially practical for the investors of movies. The other one is geographical proximity to other production companies. Over 300 film productions are located in the Gangnam area. (Film producer C)

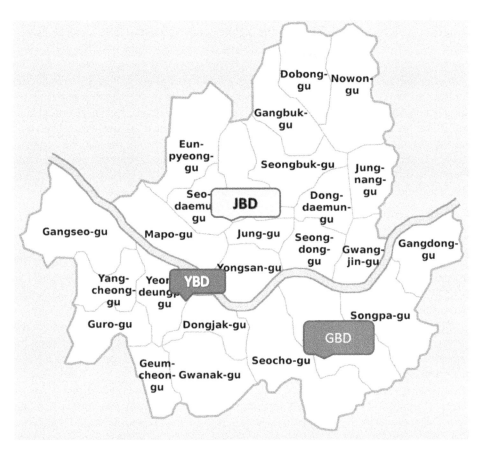

Figure 4. The three CBD in Seoul. Draft by the author.
Source: Based on Wikipedia (benutzer:ralf roletschek), http://commons.wikimedia.org/wiki/
File:Map_Seoul_districts_de.png

YBD surrounds the National Assembly and Seoul International Finance Center. Moreover, it includes bank headquarters, financial institutions and asset management companies, as well as government financial agencies, such as the Korea Exchange and Financial Supervisory Service. YBD is also the location of the three main broadcasting systems –KBS, MBC (Munhwa Broadcasting Corporation) and SBS (Seoul Seoul Broadcasting System). The Korean broadcasting industry, which was dominated by two terrestrial stations, KBS and MBC (Munhwa Broadcasting Corporation), was reshaped with the foundation of Seoul Broadcasting System (SBS) in 1991. Furthermore in 1995, cable and satellite broadcasting services began, increasing the intensity of the competition. With the greater competitive pressure, the quality of the TV products improved, increasing the popularity of TV programmes both in the domestic and later in the foreign markets as well. There are 538 TV production companies in Seoul and 291 of them are located in YBD. (Kwon, 2011, p. 73, Hwang, 2009)

Recently, a total of 200 independent TV production companies have become located in this area.

> YBD is the symbol of Korean broadcasting system history. During the 1980s, the TV productions concentrated in this area. KBS is still the center of the TV production cluster. (Korean Broadcasting System director D)

GDB and YBD, as centres of film and TV production, are successful creative clusters like those described by Hartley *et al.* (2012, p. 19). Tacit knowledge, trust relationship, localized networks, internal information flows, "know-how" and "know-who" are incentives for many film and TV production companies to stay in GBD or YBD.

Recently, the Gyeong-gi provincial government has been attempting to attract film and TV productions to relocate to the region. As part of the leading strategic industry policy of the central government enacted 2009 in order to secure sustainable job creation in the Gyeong-gi metropolitan area, a new broadcasting system cluster is being newly developed in Goyang city (Figure 5). It provides 3D computer graphic studios and aqua studios. Goyang Aqua Studio started under the authorization of Goyang city and opened as a filming studio in June 2011. It is built on an area of 25,000 square metres and is specialized in the filming of water-related scenes with four types of water pools as supporting facilities. It has attracted a large number of film projects, such as movies, TV series and TV formats including blockbusters.

Today two major broadcasting systems, MBC and SBS, are located in Goyang city. MBC established its digital production centre there in 2007 and SBS in 1995. Moreover, 86 related companies are concentrated in this area. It was not only an initiative of the Gyeong-gi provincial government to organize Goyang broadcasting system cluster, but also other demands from the company side—such as cost reduction and securing sufficient workspace.

> As a subcontract producer, my company has to locate close to MBC or SBS. It is not only physical proximity but also about informal network. In the Goyang broadcasting complex, I often meet many colleagues and contractors and we organize several informal meetings. The core information is exchanged often during this informal meeting. In our field, personal connection and proximity are still quite important. (Producing director of Subcontractor Company E)

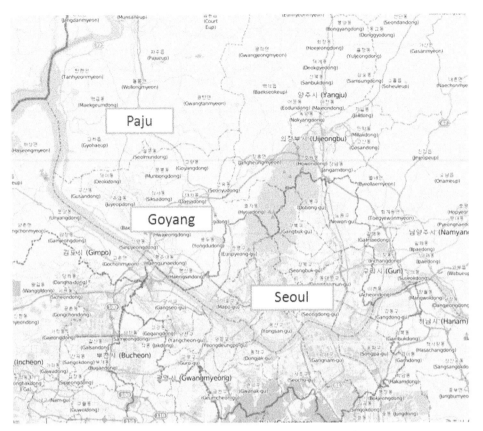

Figure 5. Dispersion of the film and TV industries in Korea.
Source: © OpenStreetMap contributors.

> We moved from YBD to Goyang city in 2007. In early 2006, the central government distributed tax deductions and subsidy incentives on condition that our company move to the Goyang broadcasting system complex. The decision making went smoothly, although some employees hesitated to commute from Seoul to Goyang. (General Manager of TV documentary production company F)

In addition to Goyang, Paju city is attractive to film productions as it offers sufficient space and relatively low rent for the post-production studios. "Heyri cultural art village", situated in Paju City, attracted scenario writers and movie technicians, in particular, in the late 1990s. At first, it was designed as "the book village" linked to Paju Publishing Town in 1997. Increasingly, lots of artists in various creative fields joined. In 2012, more than 380 writers and technicians worked in this area.

> Our company is specialized in VFX (visual effects) and CG (computer graphics). A systematic linkage with our partners is the biggest advantage of locating in Paju -Heyri cultural art village. Moreover, we are located about 50 kilometers from central Seoul; that takes only 30 minutes by car. We have here sufficient work space and a tightly -organized network.

Paju also offers artistic spaces, houses, work rooms, museums and galleries. It is definitely a "one-stop" system for movie producers. (Post-producing engineer G)

In summary, the film and TV industries are concentrated in specific districts in Seoul, namely GBD and YBD. GBD represents the centre of the film production companies and YBD the central broadcasting systems. However, recent studies (Choo, 2006; Kim, In-Hwan, 2012) not only show the concentration of the film industries within Seoul, but also reveal the dispersion of the film industries outside of Seoul. The increase in rental prices in Seoul was one of the most important motivations for many production companies that moved out of GBD and YBD, heading to Goyang or Paju, Gyeong-gi Province. The need for closer co-operation between production and post-production companies also played an important role.

3.3. *Match and Mismatch*

The concept of co-evolution is helpful for explaining the dynamics of the Korean film and TV industries as it takes the industries and their institutional settings—such as different levels of markets, governances and regulations—and interactions between them into account. The development of the Korean film and broadcasting industries reveals that their significant growth cannot be explained without considering the role of the government in supporting them (Keane, 2006; Kwon & Kim, 2013). In order to understand the current film and TV industries in Korea, it is important to look at the institutional settings and the interactions taking place among different public sectors and individual participants, as well as the discourses articulated by diverse agencies (Shin & Stevens, 2013).

Private sector firms in the film and broadcasting industries have been assisted by governments in raising capital. In the early 1990s, the Korean government removed the restrictions on the involvement of chaebols in the film, music and broadcasting industries. As a result, major chaebols—including Samsung, Lotte, Hyundai, LG and SK—entered multiple sectors of the film and broadcasting industries by acquiring a number of major companies, such as management firms, productions and distributors. In 2013 CJ E&M (Entertainment & Media), South Korea's largest entertainment conglomerate, accounted for the biggest market share of all the distributors in the domestic film market, above Lotte Entertainment and Showbox.

We strive to strengthen our market leadership in terms of production and distribution, as well as exhibition. Overall, the Korean film industry will probably be managed by a few major investors and distributors, who will also be fierce competitors in a critical situation characterized by excessive demands for strictly limited funds. (CJ E&M marketing director M)

The Ministry of Culture, Sports and Tourism is responsible for the main support policy for the creative industry. KOCCA and KOFIC are two main sub institutions supporting the film and broadcasting industries in Korea.[2] KOCCA, a government-funded agency, is a public agency dedicated to promoting Korean creative industries. It specializes in a variety of creative industries, including broadcasting, animation and game industries. The KOFIC is a special organization, entrusted by the Ministry of Culture, Sports and Tourism with the support and promotion of Korean films. KOFIC provides support programmes for Korean films such as the independent film support programme, the labour costs support programme and the accumulative support programme for released Korean films. Since July 2007, KOFIC has raised and managed the Film Development Fund, which amounts to approximately USD 430 million, in order to promote and support the Korean film industry.

> We (as a public agency) provide several funds for the independent film productions. We also support individual productions for entering overseas markets by offering diverse consulting services. (Team director of KOFIC N)

However, the programmes and budgets for the creative industries of those institutions are all linked to the changing national policy and paradigm. Often, many support plans had to be amended due to a change of government. The effect of the support policy of government agencies was thus rather volatile.

Moreover, the headquarters of KOFIC and KOCCA will be relocated to Busan and Naju by the end of 2013, following the public sector relocation policy of the central government's balanced regional development plan. The balanced regional development plan was initiated by the administration of president Roh Moo Hyun in 2003 and aimed at relocating about 40,000 public sector jobs from the Seoul Metropolitan Region (SMR) to non-SMR regions. Many internal experts reacted positively to KOFICs relocation, as it would be located right next to the Busan Cinema Centre, which is an official venue of the Busan International Film Festival.

KOCCA's relocation, on the other hand, dismays many in the field of the broadcasting industries. Naju is located in the southeastern part of Korea and has neither sufficient physical infrastructure nor human resource pools. Moreover, many interviewed broadcasting companies answered that they do not have any relocation plan involving moving out of the Seoul metropolitan area in the near future. Government-led relocation of public agencies may bring isolation and less cooperation between public agencies and creative industries, as they expect close partnership between industry and the governmental agencies.

Another mismatch problem derives from the Free Trade Agreement (FTA) between Korea and the US. When Hallyu started in the beginning of 2000, the relationship between the Korean film and TV industries and their institutional settings was cooperative. The Korean government established public agencies—such as KOCCA and KOICA—in order to support the expansion of the film and TV industries. Moreover, the central government provided diverse subsidies targeted at the film and TV industries. However, the Korean government signed an FTA with the US in 2006. The agreement includes a condition that the Korean government reduce the Screen Quota (movie screening days per year of Korean domestic films required by law) from 146 days to 73 days beginning from 1 July 2006. This was a result of continuous pressure exercised by Hollywood on the US trade negotiators, who consistently refused to enter into FTA negotiations with Korea unless the quota system was properly modified. Due to this concession made by the Korean government, both countries could launch FTA talks, and the talks were successfully concluded on 2 April 2007. They drew strong protests from the film industry. The domestic film community stressed the importance of cultural sovereignty in the film industry. In addition, many actors, actresses and directors tried to convince the Korean government to protect the screen quota system through demonstrations. Three years after the reduction of the screen quota, the average number of total admissions of

Korean films decreased to 640,109.9123 (2007–2009) from 1,107,217.82 (2003–2005) for the three years before the cutback. This reveals that the reduction of the screen quota has had negative effects for Korean films (Korean Film Council, 2012).

Recently, there has been a severe conflict between the film and TV industries and the government with regard to media censorship. The Korea Communications Commission (KCC) is a government agency that regulates TV contents. Former president Lee Myung-Bak strengthened KCC's censorship. Under the Lee administration, around 160 TV channels and journals were penalized. Protests amongst major broadcasting systems—MBS, KBS and SBS—rose in early 2012 against the excessive media censorship of the central government.

In short, in the early 2000s when Hallyu started, many institutions, whether private or public, shared a common objective, the promotion of the Korean film and broadcasting industries. Nevertheless the relationship between the film and TV industries and their institutional settings has become unstable since 2006.

4. Conclusion

This paper on the Korean film and TV industries analyses whether the notion of co-evolution can help to analyse and explain creative cluster evolution. The Hallyu phenomenon refers to the significant increase in popularity of Korean culture goods, including television programmes, films, pop music and computer games. Hallyu sparked a tremendous growth of the film and TV industries in Korea. Few studies on the Hallyu phenomenon, however, have focused on the spatial evolution of the film and TV industries. As co-evolutionary thinking indicates that organizational populations are mutually interdependent and have a reciprocal influence on each other (Schamp, 2010), it can be used to analyse and explain dynamics—for instance, localization, clustering and dispersal growth processes—in creative industries.

Based on the 35 semi-structured interviews with key stakeholders of the film and TV industries in Korea, this paper can draw three conclusions. First, Hallyu began with the export of Korean TV drama series across East Asia and the state deregulation and neoliberal reforms during the 1990s in Korea, triggering an explosion of exports of the Korean film and TV industries. Second, this development started in GBD and YBD, the core of the film and TV production within Seoul, while dispersion of those industries occurred in Gyeong-gi province. Nowadays the film and TV cluster in Seoul forms a circle with GBD as its centre, but it also includes YBD, Paju and Goyang. Third, from an institutional perspective, tensions between the central government and the film and TV industries can be observed, which have been intensifying since 2006.

In summary, the concept of co-evolution proves to be of great value in analysing and explaining the dynamics in creative industries. So far, however, the co-evolution-related literature has neglected critical events that can play an important role in changing the relations between industry/firms and the institutional environment on several spatial scales. Considering the rarity and potential of critical events, there can be little doubt about the desirability of further studies tackling the following research questions:

- Do critical events play a role as an external shock that causes certain changes in the locational pattern of the creative industries in emerging economies?

- Do creative industries in emerging economies and developed economies follow different paths?
- If yes, what role do spin-off creation, labour mobility, inter-firm networks and the creative class (Florida, 2002) play in the spatial dynamics of creative industries?

The author believes the analysis of cluster evolution provides a promising and challenging research agenda in EEG for the years to come.

Acknowledgements

An earlier version of this paper has been presented at the Regional Studies Association European Conference, Tampere, 2013. I have benefited from the useful comments made by the participants of this event and particularly by three anonymous reviewers. I would like to thank the interviewees for sparing their time. The usual disclaimer applies, however.

Funding

This research was supported by the European Science Foundation and the German Research Foundation (DFG) in the framework of the Cluster Life Cycles Project.

Notes

1. However, not all interviews are cited in this article owing to repetitions of the contents and to the limits of space.
2. There are also Korea Broadcasting Agency and Film Academy, however, they have little influence on the film and TV industry due to their weak financial capacity and instable policies.

References

Banks, J. & Potts, J. (2010) Co-creating games: A co-evolutionary analysis, *New Media & Society*, 12(2), pp. 253–270.

Berg, S.-H. & Hassink, R. (forthcoming) Creative industries from an evolutionary perspective: A critical literature review, *Geography Compass*. doi:10.1111/gec3.12156

Boschma, R. A. & Frenken, K. (2006) Why is economic geography not an evolutionary science? Towards an evolutionary economic geography, *Journal of economic geography*, 6(3), pp. 273–302.

Boschma, R. & Frenken, K. (2009) Some notes on institutions in evolutionary economic geography, *Economic Geography*, 85(2), pp. 151–158.

Boschma, R. & Frenken, K. (2011) The emerging empirics of evolutionary economic geography, *Journal of Economic Geography*, 11(2), pp. 295–307.

Boschma, R. & Martin, R. (Eds) (2010) *Handbook of Evolutionary Economic Geography* (Cheltenham: Edward Elgar).

Cho, Hae-Joang. (2005) Reading the "Korean wave" as a sign of global shift, *Korea Journal*, 45(4), pp. 147–182.

Cho, W. S. (2012) Riding the Korean wave from 'Gangnam style' to global recognition, *Global Asia*, 7(3), pp. 35–39.

Choo, S. (2006) Development of the Korean film industry and its spatial characteristics: Gangnam region of Seoul as a new cluster in a new renaissance? *Journal of the Korean Geographical Society*, 41(3), pp. 245–266.

Dahlström, M. & Hermelin, B. (2007) Creative industries, spatiality and flexibility: The example of film production, *Norsk Geografisk Tidsskrift-Norwegian Journal of Geography*, 61(3), pp. 111–121.

Florida, R. (2002) The economic geography of talent, *Annals of the Association of American geographers*, 92(4), pp. 743–755.

Gertler, M. S. (2004) *Manufacturing culture: The institutional geography of industrial practice: The institutional geography of industrial practice* (Oxford, UK: Oxford University Press).

Hartley, J., Potts, J., Flew, T., Cunningham, S., Keane, M. & Banks, J. (Eds) (2012) *Key concepts in creative industries* (London, UK: Sage).

Hogarth, H. K. K. (2013) The Korean wave: an Asian reaction to western-dominated globalization, in: *Dystopia and Global Rebellion. Eleventh Annual Conference of the Global Studies Association of North America, May 3–6, Victoria, British Columbia, Canada.* (Vol. 12, No. 1/2, pp. 135–151). Brill: Academic Publishers.

Hwang, Ki-Sung. (2009) Development plan for the fim and TV drama production in Seoul, *Seoul Film Commission.* Available at http://www.seoulfc.or.kr/ (accessed 30 January 2014).

Iammarino, S. & McCann, P. (2010) *The Relationship Between Multinational Firms and Innovative Clusters* (Cheltenham, UK: Edward Elgar Publishing).

Jang, G.-J. & Paik, W.-K. (2012) Korean wave as tool for Korea's new cultural diplomacy, *Advances in Applied Sociology*, 2(3), pp. 196–202.

Jin, D.-Y. (2012) The new Korean wave in the creative industry: Hallyu 2.0, *University of Michigan journal*, 2, pp. 3–7.

Keane, M. (2006) Once were peripheral: Creating media capacity in East Asia, *Media, culture and society*, 28(6), 835–855.

Kim, E.-S. (2012) *Korean Cinema 2012* (Seoul: Korean Film Council (KOFIC)).

Kim, In-Hwan. (2012) Goyang Broadmax 2.0 (Korea: Goyang Industry Promotion Ageny Insight). Available at http://www.gipa.or.kr/data (accessed 28 October 2013).

Koehler, R. (2013) More than just k-pop, *Korea*, 9(2), pp. 6–11.

Korean Creative Contents Agency (2001) Korean contents industry year book (Seoul, Korea). Available at www.kocca.or.kr (accessed 20 February 2014).

Korean Creative Contents Agency (2002) Korean contents industry year book (Seoul, Korea). Available at www.kocca.or.kr (accessed 20 February 2014).

Korean Creative Contents Agency (2004) Korean contents industry year book (Seoul, Korea). Available at www.kocca.or.kr (accessed 20 February 2014).

Korean Creative Contents Agency (2005) Korean contents industry year book (Seoul, Korea). Available at www.kocca.or.kr (accessed 20 February 2014).

Korean Creative Contents Agency (2013) Korean contents industry year book 2012 (Seoul, Korea). Available at www.kocca.or.kr (accessed 20 February 2014).

Korean Film Council (2003) Korean film industry year book (Seoul, Korea). Available at www.kofic.or.kr (accessed 21 February 2014).

Korean Film Council (2005) Korean film industry year book (Seoul, Korea). Available at www.kofic.or.kr (accessed 21 February 2014).

Korean Film Council (2008) Korean film industry year book (Seoul, Korea). Available at www.kofic.or.kr (accessed 21 February 2014).

Korean Film Council (2012) Korean Cinema 2012 (Seoul).

Korean Film Council (2013) Korean Cinema 2013 (Seoul, Korea).

Kwon, Seung-Ho (2011) *The Korean wave and cultural industry: A historical and political economic perspective*, Working paper of Korea Research Institute, University of New South Wales, Australia.

Kwon, S.-H. & Kim, J. (2013) The cultural industry policies of the Korean government and the Korean Wave, *International Journal of Cultural Policy*, 20(4), pp. 422–439. Available at http://dx.doi.org/10.1080/10286632.2013.829052 (accessed 28 October 2013).

Lorenzen, M. & Mudambi, R. (2012) Clusters, connectivity and catch-up: Bollywood and Bangalore in the global economy, *Journal of Economic Geography*, 13(3), pp. 501–534.

Malerba, F. (2006) Innovation and the evolution of industries, *Journal of Evolutionary Economics*, 16(1/2), pp. 3–23.

Malmberg, A. & Maskell, P. (2010) An evolutionary approach to localized learning and spatial clustering, in: R. Boschma & R. Martin (Eds) *The Handbook of Evolutionary Economic Geography*, pp. 391–405 (Cheltenham, UK: Edward Elgar Publishing).

Martin, R. & Sunley, P. (2006) Path dependence and regional economic evolution, *Journal of economic geography*, 6(4), pp. 395–437.

Menzel, M.-P. & Fornahl, D. (2010) Cluster life cycles-dimensions and rationales of cluster evolution, *Industrial and Corporate Change*, 19(1), pp. 205–238.

Miller, T., Govil, N., McMurria, J., Maxwell, R. & Wang, T. (2005) *Global Hollywood 2* (London: British Film Institute).

Ministry of Culture, Sport and Tourism (2013) *Korean Content Industry Year Book 2012* (Seoul: Ministry of Culture, Sport and Tourism).

Murmann, J. P. (2003) *Knowledge and Competitive Advantage: The Co-Evolution of Firms, Technology, and National Institutions* (Cambridge, UK: Cambridge University Press).

Nelson, R. R. (1994) The co-evolution of technology, industrial structure, and supporting institutions, *Industrial and corporate change*, 3(1), pp. 47–63.

Nelson, R. & Winter, S. (1982) *An Evolutionary Theory of the Firm* (Cambridge, MA: Harvard University Press).

Rantisi, N. M. (2004) The ascendance of New York fashion, *International Journal of Urban and Regional Research*, 28(1), pp. 86–106.

Ryoo, W. (2009) Globalization, or the logic of cultural hybridization: The case of the Korean wave, *Asian Journal of Communication*, 19(2), pp. 137–151.

Ryu, S.-H. (2013) Power entertainment Korea 2013 (Korea: KAIST). Available at http://imrc.kaist.ac.kr (accessed 23 October 2013).

Schamp, E. W. (2010) On the notion of co-evolution in economic geography, in: R. Boschma & R. Martin (Eds) *Handbook of Evolutionary Economic Geography*, pp. 432–449 (Cheltenham: Edward Elgar).

Shin, C. & Stringer, J. (2005) *New Korean Cinema* (New York: New York University Press).

Shin, H. & Stevens, Q. (2013) How culture and economy meet in South Korea: The politics of cultural economy in culture-led urban regeneration, *International Journal of Urban and Regional Research*, 37(5), pp. 1707–1723.

Ter Wal, A. L. & Boschma, R. (2011) Co-evolution of firms, industries and networks in space, *Regional Studies*, 45(7), pp. 919–933.

Institutional Context and Cluster Emergence: The Biogas Industry in Southern Sweden

HANNA MARTIN* & LARS COENEN*,**

*Center for Innovation, Research and Competence in the Learning Economy (CIRCLE), Lund University, Lund, Sweden, **Nordic Institute for Studies in Innovation, Research and Education (NIFU), Oslo, Norway

ABSTRACT *According to some scholars in evolutionary economic geography (EEG), the role of (territory-specific) institutions is relatively small for explaining where a new industry emerges and grows as firms develop routines in a path-dependent and idiosyncratic manner. This article evaluates this assertion by studying the evolution of the biogas industry in the region of Scania in Southern Sweden. The biogas is predominantly used as a fuel in the regional transport system and is considered as a crucial means to achieve environmental goals in the region. Recently, regional public policy has been actively promoting this biogas industry, aiming for cluster development. Drawing on literature from EEG and technological innovation systems, this article seeks to unpack the evolutionary process that has led to the emergence of this industry. In particular, it studies to what extent territory-specific institutions have been crucial in that respect. The analysis is case-based, drawing predominantly on in-depth interviews with key stakeholders and firms in the industry. By doing so, the paper seeks to make a contribution to our understanding of cluster development, considering the interplay between technology, industry dynamics and institutions.*

Introduction

Recently, with the "evolutionary turn" in economic geography (Boschma & Frenken, 2006; Boschma & Martin, 2007, 2010; Menzel & Fornahl, 2010; Martin & Sunley, 2011) there has been an increasing interest in the spatial emergence of economic phenomena, such as the origin of new industries. Having its roots in evolutionary economics, evolutionary economic geography (EEG) explains the uneven spatial distribution of economic activities and industrial structures based on the micro-level search and selection behaviour of firms understood as organizational routines. Emanating concepts such as related variety

89

and regional branching have added considerably to the economic geography literature as they supplement the weakness of established systemic approaches to innovation by emphasizing the influence of historical preconditions and path-dependencies in regional economic development (Boschma & Frenken, 2006; Uyarra, 2010).

With the development of the research field, however, some scholars have criticized the strong emphasis placed on to firm-level routines at the expense of institutions and other actors, for example the state (MacKinnon *et al.*, 2009; Morgan, 2012). Due to this bias, EEG has until now only rendered limited explanatory power to factors such as policy interventions and institutions in actively favouring certain development paths (Asheim *et al.*, 2013), in spite of some notable exceptions (Martin & Sunley, 2006). According to pioneering proponents of EEG, the role of (territory-specific) institutions is considered relatively small to explain where new industry emerges and grows (Boschma & Frenken, 2009). At the same time, others have argued that there is a need in economic geography to better understand institutional evolution over time with regard to regional economic change (Gertler, 2010) as there is still a rather limited understanding of the role of public policy for the diversification of regions into new growth paths over time (Asheim *et al.*, 2011).

The objective of this paper is to make explicit and specify the role of institutions for industry formation and thereby, to contribute to our understanding of cluster emergence and development considering the interplay between technology, industry dynamics and institutions. To do so, the paper studies the evolution of the biogas industry in the region of Scania in Southern Sweden. Triggered by policy programmes targeting local initiatives to reduce greenhouse gas emissions as well as experiences and existing infrastructures related to the extraction and distribution of natural gas in the region, biogas activities started to emerge in the late 1990s and early 2000s. During the past decade and simultaneously with technological advances in the biogas area, regional policies have induced further momentum of this industry, for example through creating demand for locally produced biogas by setting up environmental goals that stimulate its use in the regional public transport system. The biogas industry constitutes today an emergent industry in the region and involves a network of both public and private actors along the value chain of biogas. Actors producing and using biogas constitute the probably most visible elements of the industry; examples are feedstock producers (such as farmers and food processors), utilities as well as energy and transportation companies. Moreover, various supporting activities such as those of manufacturers of biogas equipment are located in the region, and also local universities, research centres, support organizations and regional policy are involved in the biogas activities, explicitly aiming at strengthening cluster development. From an EEG perspective, the biogas industry in the region of Scania constitutes a relevant case to study the process of cluster emergence. In particular, this paper aims at investigating how territorial institutions, in combination with firm-level routines and technology development, can steer regional economic development and evolution along certain development paths. In doing so, institutions are primarily considered in the shape of policy interventions, including their impact on actors' behaviour. The following research questions are addressed:

> How do specific territorial institutions matter for the emergence of the biogas industry in Scania?
> How do policy interventions work actively in favour of new regional economic development paths?

The pronounced analytical focus on the role of public authorities and policies calls for taking a combined evolutionary-institutional perspective on cluster emergence and regional industrial development. Accordingly, the theoretical framework of the paper departs from a discussion on industry emergence in (evolutionary) economic geography introducing concepts such as path-dependence, related variety and regional branching. In order to account for an institutional perspective, the article draws on insights from innovation studies targeting the functions of technological innovation systems (TIS) regarding transformative technological change. The paper constitutes a theoretically informed case study, predominantly drawing on in-depth interviews with key stakeholders in the industry.

The paper is organized as follows. The next section presents the theoretical framework of the study, drawing on literature on spatial industry emergence as well as on TIS. The subsequent section introduces the empirical case study and analysis, also including an outline of the research design and methods applied in the study. The paper ends with a discussion and conclusion section.

Theoretical Framework

Spatial Industry Emergence and Evolution

Traditionally, the major part of the literature on economic geography and in particular, the subfield of geography of innovation, has been focusing on localized learning and agglomeration externalities for innovation processes. The functioning of clusters, understood as a geographical concentration of interconnected firms and various support organizations active in a particular field (Porter, 1998), as well as policies to support them has gained considerable attention within the discipline. The research done on clusters has, however, brought about a rather static view, hardly paying attention to a long-term perspective and consideration of the early roots of economic activities. Consequently, questions such as "how clusters actually become clusters" and "which factors and dynamics lead to their emergence" have hardly been addressed (Menzel & Fornahl, 2010).

The emphasis on the evolutionary nature of innovation processes in EEG responds consequently to an important critique raised against the "traditional" literature on the geography of innovation in that it provides snapshots of successful regions detached from their time–space context (MacKinnon *et al.*, 2002; Shin & Hassink, 2011). According to Uyarra (2010) the majority of regional innovation system studies can be characterized as "inventory-like descriptions of regional systems, with a tendency to focus on a static landscape of actors and institutions" (p. 129). Having its roots in evolutionary economics, the notion of path-dependence is central to EEG, denoting the importance of history and the dependence of past decisions for future events to occur. As a consequence, it is a widely shared understanding in the field of EEG that regional economic development is path-dependent. As Martin (2010) frames it, it is "the combination of historical contingency and the emergence of self-reinforcing effects" stemming from critical mass and spillovers that is considered key in steering the "technology, industry or regional economy along one 'path' rather than another" (Martin, 2010, p. 3).

Initial attempts in economic geography to understand industry emergence by paying attention to path-dependencies in regional economic development followed the "window of locational opportunity" (WLO) line of thought (Scott & Storper, 1987;

Storper & Walker, 1989). This literature argues that new industries experience a rather high degree of locational freedom as they put relatively novel demands on their locational conditions in terms of access to knowledge, labour skills and machines. As these requirements are still uncertain and not in place yet when a new industry starts to form, all regions have a similar potential to become the host of a new industry (Boschma, 1997). Once a critical number of firms carrying out a new type of industrial activity has established itself in the region, the WLO narrows down because the industry becomes tied to its location. In this manner, an industry becomes locked-in in a specific place (Storper & Walker, 1989). With regard to industry emergence, the WLO model assumes that new industries form and shape regional economic spaces (rather than the other way around) and ascribes much explanatory power to the role of chance and accidental events (Nygaard Tanner, 2012), thus resonating with the traditional model of technological path-dependence as laid out by David (1985) and its emphasis on "historical accidents", "chance events" or "random" action for new technological pathways.

With the evolutionary turn, however, voices have recently been raised for a re-interpretation of path-dependencies, implying a stronger consideration of local (knowledge) resources in shaping regional industrial development paths over time (Martin & Sunley, 2006; Trippl & Otto, 2009; Simmie, 2012; Strambach & Klement, 2013). The literature on EEG gives evidence to path-dependent regional development, stating that firms are expected to diversify into activities that are technologically related to their existing competences. Consequently, regions are assumed to slowly diversify and branch out into technologically related fields, implying that industrial structures are rather persistent in a region (Boschma & Frenken, 2011a; Boschma & Martin, 2010). This industrial development and evolution is explained from an EEG perspective by knowledge spillovers between firms, assuming that for effective learning to take place a certain degree of cognitive proximity (or technological relatedness) between firms is needed so that firms can interpret, absorb and implement new knowledge (Cohen & Levinthal, 1990); however, also a certain degree of cognitive distance between actors is needed to stimulate novelty (Nooteboom, 2000). To address the question of optimal cognitive distance in a context of knowledge spillovers at the regional level the concept of "related variety" has been introduced (Frenken et al., 2007), stating the positive impact of a variety of different yet technologically related regional industries on regional growth.

Due to its roots in evolutionary economics, the EEG framework has a pronounced perspective on, and interest in, firms and their routines. More precisely, the pioneering work on EEG (Boschma & Frenken, 2006) makes an explicit distinction between evolutionary and institutional approaches to economic geography, arguing that the role of (territory-specific) institutions is relatively small to explain where a new industry emerges due to the fact that firms develop routines in a path-dependent and idiosyncratic manner (Boschma & Frenken, 2009). This work does not neglect the impact that (territorial) institutions can have on the behaviour of firms, but institutions are treated as conditioning rather than determining the behaviour of firms and regional development as a whole (Boschma & Frenken, 2011b). Moreover, it is argued that institutions come into existence or become aligned to support a specific industrial activity once it has started to develop (Boschma & Frenken, 2009). As such, EEG follows the general line of arguments laid out in the WLO model in assuming that institutions are responsive to, rather than responsible for, new development paths.

The mentioned work on EEG has led to a general understanding in the discipline of economic geography that regional economic development is not random but that it relies on historical prerequisites in terms of firms' knowledge bases and routines as well as knowledge spillovers that lead to new industry emergence over time. However, the fact that the pioneering work in EEG puts much emphasis on path-dependencies in regional economic development has been taken up by scholars in the literature, arguing for an incorporation of institutions in approaches to explain path-dependence as well as a stronger consideration of change processes in evolutionary thinking (Martin, 2010) or emphasizing the importance of processes of collective agency in creating and steering certain development paths (Simmie, 2012). Others have mentioned their concern about a "theoretical relegation" of institutions and social agency (MacKinnon *et al.*, 2009), while some such as Essletzbichler (2009) and Grabher (2009) regard a stronger consideration and inclusion of institutions in EEG as highly relevant for the further development of the research field (Asheim *et al.*, 2013). As such, there is still a limited empirical and theoretical understanding of the role of institutions and public policy concerning the diversification of regions into new growth paths over time (Asheim *et al.*, 2011), as well as a lack of scientific work taking a more holistic perspective regarding the co-evolution of institutions and technology (Strambach, 2010).

Institutional Context and Industry Formation

In contrast to the literature on EEG, the literature on TIS and, more broadly, socio-technical transitions, allows taking a co-evolutionary perspective on technology and industry dynamics and their institutional embedding. A core tenet of this literature is that technology and institutional dimensions should not be analysed separately when trying to understand innovation. Rather, both aspects are understood in their co-determination over time. The analysis is therefore not restricted to "technologies" but rather addresses "socio-technical systems". The formation of socio-technical systems is conceived as a process of constructing "configurations that work" (Rip & Kemp, 1998) among technological artifacts and their organizational, institutional, infrastructural and use related aspects. During early formation phases largely all major components of a socio-technical configuration are still in flux: technologies need to improve in performance and cost characteristics, use patterns and user preferences have not yet been fully established and institutions to regulate the impacts of the technology are not yet fully spelled out (Dosi, 1982; Callon, 1998). On the other hand, established and mature socio-technical configurations may exhibit strong path-dependencies that go beyond lock-in effects based on increasing economies of scale (Arthur, 1994), but may be generated by the initial establishment of use patterns (David, 1985), standards, infrastructures or institutional structures (Granovetter & MacGuire, 1998).

Within the literature on socio-technical transitions, the TIS approach, introduced by Carlsson and Stankiewicz (1991) and further developed by among others Hekkert *et al.* (2007), Bergek *et al.* (2008) and Markard and Truffer (2008), has gained considerable attention in developing a process view on early industry formation based on emergent technological fields. The framework takes a systemic perspective on innovation and considers different actors such as governmental and non-governmental organizations, research institutes and firms as well as different forms of institutions and their interplay as important elements for innovation to take place. Following Markard and Truffer (2008), a TIS can be defined as "a set of networks of actors and institutions that jointly interact in a specific technological field and contribute to the generation, diffusion and utilization of variants of a new technology and/or a new product" (Markard & Truffer, 2008, p. 611). As a framework for analysis, the TIS approach has a rather strong focus on mapping the functionality of the innovation system. In order to assess the performance of the innovation system, Johnson and Jacobsson (2001) and Bergek *et al.* (2008) have identified seven functions that have to exist around a new, emerging technology (i.e. they have to be carried out by actors and institutions) in order for a technology to diffuse and to lead to new industry emergence: (1) *knowledge development and diffusion* (generation, diffusion and combination of knowledge in the innovation system), (2) *influence of the direction of search* (incentives for organizations to enter the TIS), (3) *entrepreneurial experimentation* (reducing uncertainty through probing and bringing a technology into practice), (4) *market formation* (development of markets for emerging technologies), (5) *resource mobilization* (mobilization of financial and human capital), (6) *legitimation* (exert influence on the public opinion with regard to a new technology) and (7) *development of positive externalities* (achievement of clustering effects in the emerging industry) (Bergek *et al.*, 2008).

The strong focus on functions in the TIS framework has brought about important insights with regard to key activities in innovation systems as well as understanding processes of technological change and innovation (Hekkert *et al.*, 2007). This allows making statements concerning an active construction and the set-up of a supportive institutional context in emerging (clean-tech) industries. Furthermore, the TIS framework makes it possible to take a dynamic systems perspective on innovation regarding how specific functions have come in place. The TIS functions target various networks, actors and institutions of the system and makes it obvious that for new technologies to penetrate markets, multiple dimensions play important roles. The core strength of the framework on mapping the functionality of innovation systems as well as its underlying strength concerning policy implications to support new technologies is yet accompanied by its weakness in explaining regional differences in technology evolution and development (Coenen *et al.*, 2012).

Due to the strong focus of EEG on firms and their routines as main protagonists, the TIS framework will thus act as a complementary perspective to help specify the role of institutions on the early emergence process of the biogas industry in Scania.

Analysis

The biogas industry is considered to constitute an emergent industry in Scania, a region that traditionally has been characterized by its agriculture and food industries (producers of organic waste). Today, the region hosts a broad network of public and private actors on the supply as well as demand side of biogas. Figure 1 illustrates that the biogas activities in Scania cover the entire value chain, including feedstock production, collection and transport, pre-treatment and upgrading of biogas, distribution and retail as well as end-use (Ericsson *et al.*, 2013). Furthermore, it becomes apparent that the value chain actors represent existent sectors and industries (i.e. agriculture and food industry as well as the energy and waste sector) that become connected to one another through the value added created with biogas. In addition to these activities that are directly integrated into the value chain, supporting activities such as those of various producers of biogas equipment,

VALUE CHAIN	VALUE CHAIN ACTORS	SUPPORTING ACTIVITIES
Feedstock production ↓	Farmers Industrial food processors Households Municipal waste water treatment Service sector	Waste collection equipment
Collection and transport ↓	Farmers Municipal waste management Municipal waste water treatment	Storage of substrate
Pretreatment and upgrading ↓	Energy companies Farmers Municipal waste management Municipal waste water treatment	Biogas upgrading equipment Gas engines and gas turbines Plant engineering companies Pretreatment equipment
Distribution and retail ↓	Energy companies Oil companies Municipal waste management Municipal waste water treatment	
End-use	Private gas vehicle owners Public transport, municipal vehicle fleets	

Figure 1. Value chain actors of biogas and supporting activities in Scania.
Source: Own illustration based on Ericsson *et al.* (2013).

to a large part also having their roots in the direct value chain activities, are located in the region.

Based on Ericsson *et al.* (2013), a number of ca. 40 companies[1] can be identified in Scania that are either directly or indirectly part of the biogas value chain, not yet including actors in the service sector, farmers, private vehicle owners, universities, research centres, consultants and cluster initiatives. As shown in Table 1, Scania (Skåne) is today the county with the highest biogas production as well as count of biogas plants in Sweden, producing ca. 0.3 TWh of energy (in 2011) and amounting ca. 20% of the Swedish overall biogas production.

Scania is aiming for an increase of biogas production to 3 TWh in 2020, equal to 10% of the county's total energy demand, and by doing so, to create 3300 new jobs in the region (Region Skåne, 2011). Furthermore, the region wants to develop into an internationally leading "Centre of Excellence" for biogas, that is a collective term used for establishing internationally leading research and collaborations within the biogas field.

The following sections will shed light on the formation process of the described industry. Due to the fact that biogas-related activities do not underlie any industry classification code, it is not possible to make statements on the development and (regional economic) impact of that industry in quantitative terms; that is in terms of employment or turnover. The analysis is therefore based on a combination of qualitative research methods, with personal semi-structured in-depth interviews constituting the main data source. The inter-

Table 1. Biogas plants and biogas production in the Swedish counties 2011

	Plants (count)	Production	
		County (GWh)	% of national total
Blekinge	3	1.4	0.1
Dalarna	12	24	1.6
Gotland	1	7.3	0.5
Gävleborg	6	11.7	0.8
Halland	11	72.7	4.9
Jämtland	3	9.2	0.6
Jönköping	11	39.6	2.7
Kalmar	10	28.4	1.9
Kronoberg	6	10.2	0.7
Norrbotten	8	32.5	2.2
Skåne	41	289.8	19.7
Stockholm	16	257.1	17.5
Södermannland	7	51.8	3.5
Uppsala	5	34.8	2.4
Värmland	10	14.5	1.0
Västerbotten	5	36	2.4
Västernorrland	13	109	7.4
Västmanland	8	43.5	3.0
Västra Götaland	35	215	14.6
Örebro	11	66.9	4.5
Östergötland	11	117.5	8.0

Source: Energimyndigheten (2012), own table.

views are based on a thematically structured guide containing pre-formulated central questions about the interview content. In this way, semi-structured interviews ensure a certain level of comparability between the single interviews, while at the same time allowing for individual adjustment to the experiences of the respective interviewee. As such, the semi-structured interviews have proved to be a suitable method for the explorative, however theoretically informed, design of the study. Additionally, the interview findings are complemented by document studies on publicly available data sources such as strategy documents and annual reports (Länsstyrelsen, 2011; Region Skåne, 2011; Energimyndigheten, 2012; Skånetrafiken, 2012). In total, the paper draws on a number of 17 interviews with key stakeholders of the biogas industry, involving public sector and industry, as well as a major university in the region. Eleven interviews were conducted between September 2012 and April 2013, explicitly addressing the research questions studied in the paper. An additional six interviews, conducted in May 2013 within the framework of a related research project, were used as reference and for cross-checking purposes. The interviews were in large part conducted in Swedish (a minority of them in English; depending on the interviewees' preferences), and transcribed and translated into English by the authors. Although the interviewees chosen for this study had different backgrounds (i.e. in the private or public sector or in research), they showed general agreement with regard to the mechanisms operative during the early phases of biogas industry development.

Emergence of the Biogas Industry: Unpacking the Evolutionary Process

Setting the scene: early activities in the region. Activities related to the production of biogas in the region of Scania had their first origin in the beginning of the 1980s and were a reaction to national regulations targeting the reduction of sludge emanating from water treatment at purification plants. Looking back, these early activities can be seen as one important element for the formation of the industry at later stages; at these times, however, these processes were also taking place in other parts of the country and moreover, they were not exclusively targeting the production of energy.

> The production began at wastewater treatment plants, in order to find ways to treat the sludge and to reduce its volume. (...) You got energy that was used to heat the houses around, but at that time it was not the main purpose to produce energy—and it happened in many parts of Sweden. (Project coordinator, stakeholder association)

Also going back to the early 1980s, simultaneous activities of a large energy company in the region concerning the supply of Southern Sweden with natural gas and the construction of a natural gas grid along the region's west coast can be considered as crucial component for the later formation of the biogas industry. These activities can be seen as a strategy targeting secured energy supply in the southern part of the country as reaction to a national nuclear power referendum in 1980 to decide on the close-down of a nuclear power plant in the region. The experiences with natural gas as a new source of energy in Southern Sweden (also closely related to the oil crisis in the 1970s) as well as progress in technology development led to the political decision in the mid-1990s to run public city traffic (i.e. the city busses) in the region's capital Malmö on natural gas in order to reduce emissions, improve urban air quality and reduce traffic noise. In other parts of the region, mostly driven by environmental concerns, early attempts were made by some municipalities to collect organic waste from households and to use it as a renewable energy source. Simultaneously in the late 1990s, the previously mentioned energy company became involved in a municipal project targeting pilot experiments concerning the feed-in of biogas into the natural gas grid. Due to this demonstration project, the energy company became an international forerunner in upgrading technologies, targeting the upgrade of biogas to natural gas quality in order to distribute it through the existing grid and making biogas available in other parts of the region. The development of upgrading technologies can, in retrospect, be considered as a crucial element for industry build-up at later stages.

> Particular for Scania was the natural gas grid as it is very limited in Sweden. That was a main reason to start the technology, to inject biogas into the natural gas grid. This development was unique in the world. (Project manager, energy company)

At that time (i.e. between 1998 and 2002), the regional capital set up local environmental goals to reduce greenhouse gas emissions and biogas was increasingly considered as a future ambition in public transport.

Industry emergence. The emergence of the biogas industry in Scania gained momentum in 2002, when the Environmental Protection Agency of Sweden (i.e. the national governmental agency responsible for proposing and implementing environmental policies) announced a so-called Climate Investment Programme (KLIMP). KLIMP constituted a seven-year grant covering the period 2002–2008 and targeted local initiatives focusing on the reduction of greenhouse gas emissions and increasing energy efficiency in Sweden.

> Previously, there was a support programme called 'KLIMP' and there one could apply for money to reduce the environmental damage. Biogas got a large share of this money because it was considered the best activity as it gives significant climate effects. But companies could not get this money, only municipalities could apply. (...) They saw that there was much that could be achieved with regard to biogas in Scania. (Project coordinator, stakeholder association)

Altogether, biogas projects in 17 of Sweden's 21 administrative regions received a grant. However, the region of Scania stood out and received together almost half of the overall grant (of a total of 622 million Swedish krona). As such, the activities and built-up of infrastructures in the region of Scania prior to KLIMP during the 1980s and 1990s can be seen as a critical factor for success with regard to the KLIMP applications. KLIMP can be regarded as an institutional setting providing legitimacy for technological change targeting increased energy efficiency and a reduction of greenhouse gas emissions by exerting influence on the general and public opinion of new technologies. The programme can be considered as an instrument to communicate the message that technological change was desirable by relevant actors. By implication, as legitimacy influences expectations, KLIMP can likewise be seen as means of guiding the direction of search of actors and creating incentives to enter the TIS (Bergek *et al.*, 2008). Hence, KLIMP provided stability and was a crucial element for steering the development along a new development path (or a technological trajectory). However, at that time only comparatively little industrial activities targeting biogas were existent in Scania. Rather, the previously mentioned simultaneous (and largely independent) activities and prerequisites in the region such as suitable infrastructure and experiences with natural gas, a demonstration project targeting biogas upgrading technologies as well as early activities to collect organic waste from households can be seen as factors constituting an anchor for KLIMP to build on. The development in Scania was further strengthened in December 2002 when the County Administrative Board of Scania published an environmental action plan containing specific milestones concerning the reduction of greenhouse gases in the region. This action plan was worked out by a large number of municipalities, organizations and increasingly also companies in the region.

The network of actors involved in biogas activities in Scania gained increasing foothold in year 2005/2006 when a regional association for biogas stakeholders (Biogas Syd) was founded, driven by various public and private biogas actors in the region.

> It was the biogas actors in the region who felt that they needed an organisation that collected all the questions and pushed them. The initial actors were waste management companies, energy companies, universities and some municipalities. (Project coordinator, stakeholder association)

The association can be seen as bottom-up initiative resulting from a growing need for operational and strategic interaction in the biogas area and it is founded and funded by its members as well as by regional authorities providing basic funding. Thereby, the increasing need for interaction was resulting from the actors' rising awareness that biogas has a high potential in Scania. First, the region is characterized by a high amount of raw material (biomass) through the region's traditional stronghold in agriculture and food industries. Second, there was an increasing interest among actors such as energy companies and utilities to develop a more environmentally friendly profile. Furthermore, and also drawing on earlier experiences with natural gas, research regarding biogas technologies had made substantial progress in the region, leading among others to spin-offs from a technological university in the region. The establishment of the stakeholder association, being a support and network organization with the aim to increase the production and use of biogas in the region, can be seen as important for strengthening the regional networks in the regional biogas field. From an institutional perspective, the foundation of the network organization can be considered as supporting knowledge development and diffusion, a function regarded as central for a TIS and innovation processes in general (Bergek *et al.*, 2008). Network activities are considered crucial for knowledge exchange and interactive learning, and in the region of Scania these clearly profited from different, but related industrial activities existing in the region, such as agriculture and food industries. Likewise, the foundation of the network organization can be seen as means of resource mobilization (both financial and human capital).

From industry emergence to growth. A decisive moment for the biogas industry in Scania was in 2007 when the regional government's public transport committee set up a goal that all public transport in the region should be fossil free in 2020, with subgoals targeting fossil free city traffic (city buses) in 2015, regional traffic in 2018 and remaining service trips in 2020. In reaction to the announcement of these goals, the company running the public transport in the region—being a publicly owned company and part of the regional authorities—thereupon took the decision to invest in biogas. Important for this decision was the fact that the energy needed for the public transport should be produced locally in order to obtain a direct environmental effect in the region. Biogas was regarded as the fuel with the highest regional potential; attributed also to the increasingly developing regional specialization in biogas.

> There was a great potential seen in Scania and it has to do with the agriculture here. There are many residual products from agriculture, Sweden's best agricultural land is in Scania. This was a very important aspect, plus that in Scania there is a strong food industry and residual waste that can be used, (...) also including residual waste from households. (Head of strategic development, transportation company)

Furthermore, it was important for the transportation company, acting on behalf of the regional government, to decide only on one technology and not on several at the same time.

> And it was also important to invest in one fuel. By focusing on biogas there was a clear signal given to the market that it is biogas that counts in Scania. (Head of strategic development, transportation company)

The latter can be seen as a clear example from the regional authorities to support entrepreneurial experimentation through reducing uncertainty and facilitating concrete actions targeting a specific technology (Hekkert *et al.*, 2007). Moreover, the grant from the KLIMP programme, still in place when the regional climate goals were set up, was used for the acquisition of biogas buses and public filling stations. The regional public transport system thus played a crucial role in promoting the development of the biogas industry in Scania as it created a local market for the biogas produced in the region. The decision taken for the regional public transport system led to activities of private companies (such as energy companies) to extend investments regarding the (commercial) production of biogas.

The development of the industry was further supported in spring 2010 when the County Administrative Board of Scania set up a climate goal for the region, particularly to bring forward the regional production and consumption of biogas. The goal implies a total production of 3 TW biogas in the region in 2020, which equals 10% of the county's energy demand. As a reaction to that, in December 2010, a roadmap was worked out by the regional government together with municipalities, the County Administrative Board of Scania, universities and private companies in order to concretize specific actions to reach the above-specified goals and to form a basis for improved co-operation between actors in the industry. Although the action plan was worked out in collaboration with private companies and research organizations, the regional government played a major role in its development. As such, the roadmap is in line with the decision on fossil free public transport taken in 2007, and can be seen as further signal providing legitimacy concerning the future support of biogas-related activities in the region. The effect of the legitimacy and market creation on the industry becomes evident from the fact that over time increasingly also private actors have been entering the industry and that the industry is diversifying in terms of markets. Whereas in 2007 public transport was almost the only—and still is the dominant—commercial consumer of biogas, it has become increasingly accepted as a biofuel among private vehicle drivers in the region also.

> Five to six years ago, Region Skåne [i.e. the regional government] was one of the major consumers of biogas, and it still is, but it has grown a lot of interest among private car drivers as well as among companies, i.e. company cars. (Head of strategic development, transportation company)

Future development. Although the biogas industry build-up has constituted a rather unproblematic process (apart from minor teething troubles with regard to the functioning of new technologies) that hardly encountered any resistance by actors, the industry may face challenges with regard to future growth. These are mainly a matter of missing explicit rules and transparency on the national level that run the risk of inhibiting the further development of biogas activities in Scania. Due to lacking long-term perspectives provided by national policy concerning regulations and incentives, actors become hesitant with regard to their investments—both on the production and demand side of biogas (Region Skåne, 2011). By way of example, recent plans of an energy company to construct a biogas plant in the region applying a new and more efficient technology were put hold on as the required investments are enormous and too uncertain in times of non-transparent regulations regarding tax benefits of biogas. As the decision of private households and companies to invest in biogas vehicles is much driven by financial incentives, the further future development of that market is unclear. The price of biofuels (and renewable resources in general) has to be seen in relation to that of fossil resources, calling for a provision of long-term legitimacy also from higher political levels.

> The price of [natural] gas is not increasing as much as it would be needed to be able to invest. (. . .) It is because [the price of biogas] cannot rise above the price of oil, petrol and diesel, but it rather has to be seen in relation to it. It may not be more expensive to run biogas—then you do not get any consumers. (Project coordinator, stakeholder association)

Discussion

As identified in the analysis, the emergence of the biogas industry in Scania was favourably conditioned by the co-location of different but related sectors encompassing largely the entire value chain for biogas production and consumption. However, the latent potential for diversification of incumbent industries and subsequent processes of branching offered through the production of biogas did only gain momentum through the announcement of the national KLIMP programme. By targeting local initiatives for energy efficiency and a reduction of greenhouse gasses, it provided much-needed legitimacy for technological change within waste and energy companies, while at the same time influencing actor's expectations and guiding their direction of search. The provision of legitimacy through KLIMP led subsequently to an arising need for increased knowledge development and diffusion, becoming apparent in the foundation of a network and support organization for biogas (Biogas Syd). This need can be seen as a path-dependent result of the preceding policy interventions, yet this time increasingly driven by private actors as a response to the legitimacy provided by the state towards (local) public authorities. Decisive for the further development of the emerging biogas industry in Scania were the subsequent political decisions taken by the regional government. By appointing biogas as a fuel with a positive future prospect, the regional authorities were tying up with the technological trajectory catalysed by KLIMP. In doing so, they became active in supporting regional market formation which can be seen as a crucial process for the commercialization of a new technology and product as it provides further legitimacy, entrepreneurial experimentation and guidance of the search for the actors.

This (stylized) account shows that the emergence of a biogas cluster in Scania was indeed partly conditioned by related variety between incumbent industries in the region as suggested by the literature on EEG. Its realization was, however, strongly influenced by policy decisions and subsequent collective action by industry actors as a response to these policy decisions. Drawing again on evolutionary thinking, it can be argued that these policy decisions were highly important in shaping a path-dependent process of inter-related technological and industrial change. The complementary TIS perspective helped specify how policy, as well as the responses by the actors, has been crucial for creating legitimacy among actors to invest and commit resources to a new technological development path. In addition, policy-led market creation for biogas further fuelled the legitimacy of this development path, encouraging firms to experiment, in a process of entrepreneurial discovery, with commercialization of biogas related infrastructure.

The fact that the TIS functions could be identified to have operated, as well as interacted with one another, during the early evolution of the biogas industry calls for a more detailed elaboration of the spatial levels at which the functions were supported. In order to allow for a spatially differentiated analysis, Figure 2 demonstrates the evolution of the biogas industry in Scania, yet in a simplified manner.[2]

First and foremost, the figure illustrates that the majority of TIS functions is supported at the regional level (i.e. in Scania) whereas only the legitimacy and the mobilization of financial resources by the KLIMP programme are realized at the national level. As KLIMP initially targeted initiatives in all Swedish municipalities, it could have led to similar effects in all Swedish regions. However, the momentum in Scania was achieved due to the existence of specific local "prerequisites" prior to KLIMP was announced. These include the energy sector, particularly technological and infrastructural experiences regarding natural gas (existence of the grid, development of upgrading technologies) as well as traditionally strong food industries and agriculture providing resources, that is residuals, for biogas. The presence of these industries should thus be regarded as decisive for the success of the KLIMP applications stemming from municipalities in Scania. It has to be noted, however, that KLIMP did not target industry emergence; rather, industry emergence should be seen as an implication and consequence of KLIMP as it provided a common orientation to previously existent, but largely independent activities. KLIMP played thus an important role in aligning the activities of actors in the region, and by implication, creating linkages between established industries. In other words, KLIMP influenced actors' direction of search and steered the development towards a new technological path through diversification of existing industries.

Based on these effects of KLIMP, several other processes (i.e. functions) such as a growing need for further knowledge development and exchange, resource mobilization, market creation, entrepreneurial experimentation and legitimacy became operative at the regional level. All these developments result either from decisions that were actively taken in the region and/or that profited substantially from the regional level. It has to be

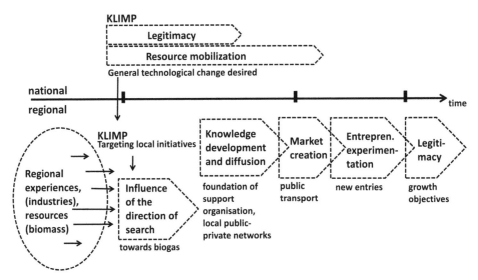

Figure 2. TIS functions—a spatial perspective on biogas development in Scania (simplified illustration).
Source: Own draft.

noted, however, that these processes do not follow any strict sequence as pictured in Figure 2; in some cases they overlap and are simultaneously active. Furthermore, not all these functions should be understood as being exclusively regional in nature as they may be linked to developments at the national or international level. By way of example, Scania is very likely embedded in knowledge networks exceeding the regional boundaries and moreover, developments targeting biogas are influenced by a general (international) endeavour to reduce the use of non-renewable resources.

In Scania, however, the regional level is considered crucial with regard to biogas industry formation. This is due to the specified regional "prerequisites" such as the presence of related industries. The process of industry diversification in the region was, however, initiated by a national policy programme (i.e. KLIMP), and would otherwise not have gained momentum; at least not at that time and not in the same manner. Subsequently, political decisions made in Scania itself led to further development of the biogas activities. This demonstrates that institutional dimensions should not be relegated in accounts of cluster evolution, especially in early stages of cluster emergence. Rather, it complements the firm- and competence focused perspective elaborated in EEG in helping explain under what conditions and how actors realize the latent potential for industry diversification.

Conclusions

The objective of this paper was to make a contribution to the understanding of cluster emergence and development from a co-evolutionary perspective involving technology, industry dynamics and institutions. To do so, the paper took a combined institutional-evolutionary perspective by, on the one hand, drawing on literature from EEG concerning path-dependence in regional economic development and the question where new industries form and why they form where they do. On the other hand, to account for the complementary institutional perspective, the paper made use of the literature on socio-technical transitions, particularly the TIS approach, concerning an active construction and the set-up of a supportive institutional context in emerging technologies and (clean-tech) industries. By studying the emergence of the biogas industry in the region of Scania in Southern Sweden, the aim of the study was to bring to light to what extent territory-specific institutions have mattered for its emergence—and in particular, to reveal how policy interventions can work actively in favour of new regional economic development paths.

Referring to the discussion on industry emergence in EEG, neither does the case support the WLO argument of industry emergence and location being a random phenomenon, nor can industry emergence be exclusively explained by firm-level routines. Rather, the analysis reveals that specific territorial institutions can (and do) matter for regional industry emergence. In Scania, a national policy programme (KLIMP) targeting technological change and energy efficiency was crucial for aligning existing competences and activities (in the broader sense) to one another which were different but not entirely unrelated with regard to biogas. In other words, infrastructures and technologies targeting natural gas proved to show synergies to biogas—and the existence of the closely related activities in agriculture and industrial food production proved to provide residuals. The potential of these synergies was, however, not made use of before KLIMP was announced. Here, the policy programme was decisive for steering the region towards a new development path.

This alignment of interests and expectations, caused by the policy programme, led subsequently to increased cooperation and a need for further knowledge development among both public and private actors in the region. Hence, it is a striking example of how policy programmes can (and do) shape behaviours of actors. A strategic perspective for the industry was thereupon built-up by active policy decisions on the innovation system in the region itself, i.e. by creating a market for the produced biogas and by setting up further strategic development goals. Thus, we argue that in the case of the biogas industry in Scania, institutions have initiated and strengthened a number of processes that were key to technological development, which, in turn, were driving the emergence of a cluster forward. These were either directly created by policy programmes, and respective decisions—or they developed as a response to these. The processes started to interact with one another and created path-dependencies and stability for technology and industry development at the regional level.

By way of concluding, we argue that the TIS framework has been helpful in making explicit and specifying the role of institutions, in the shape of public policy interventions, for the development and diffusion of new technology, which in turn gave rise to cluster emergence. These institutions explain to a considerable extent the path-dependent evolution of the cluster. While the general applicability of the patterns found in this study are limited—given the design of the study it has been impossible to control for local contingencies making broader generalization to other regions, industries and technologies troublesome[3]—we argue that the TIS framework can be fruitfully employed in the analysis of cluster emergence as it helps draw attention to crucial processes in the build-up of a (technology-based) industry. While "traditional" EEG approaches focus primarily on industrial and knowledge dynamics, we would argue that this is only part of the story. The case illustrates that suggested TIS functions such as legitimacy creation, market formation, and guiding the direction of search and entrepreneurial experimentation are indeed important processes that need to be taken into account, in addition to knowledge development and diffusion, to arrive at a more comprehensive understanding of the complexities involved in industry and cluster emergence, which includes a more prominent role of institutional factors.

This study has taken the TIS functions as a point of departure, which may to some extent limit the analysis at the expense of other processes and institutional factors that could be important in a context of cluster emergence (e.g. factors that condition the policy design of regional and national government, institutions underpinning the collaborative behaviour of firms (across sectors), factors that influence the propensity for public–private coordination). While it goes beyond the scope of this study, this article does invite future research in the context of cluster emergence and evolution not to relegate the role of institutions as this seems to be particularly important in early stages of cluster evolution. Here, we believe that insights on knowledge and industry dynamics from EEG would be highly compatible with adopting a systems approach to innovation and the central role played by institutions in this literature.

Acknowledgements

The authors would like to thank two anonymous referees and the guest editors for their helpful comments. Special thanks also go to all interviewees for their commitment as well as to Per Wretlind for providing additional interview material. The paper has received constructive feedback at the Association of American Geographers' Annual Meeting in Los Angeles, the Annual European Conference of the Regional Studies Association in Tampere, the fourth International Sustainability Transitions Conference in Zurich and the Eu-SPRI Early Career Researcher Conference in Lund, 2013. Particularly, the authors are grateful for the valuable comments of Pål Börjesson, Koen Frenken, Marko Hekkert, Hans Hellsmark, Lars J. Nilsson and Andrés Rodríguez-Pose on earlier drafts of this paper.

Funding

Financial support is gratefully acknowledged from the Cluster Life Cycles project funded by the European Science Foundation, the KIBIOS project funded by the Swedish Energy Agency and the TOP-NEST project funded by Nordic Energy Research. Moreover, this work was supported by VINNOVA [core funding grant 2013-05115] and by the Swedish Research Council [Linnaeus grant 349-2006-80].

Notes

1. Private companies and municipal actors in the region that were identified in Ericsson *et al.* (2013) are VASYD, NSV A, Kristianstad municipality (municipal waste water treatment); VASYD, Kristianstad renhållning, NSR, Sysav Biotec 2012, Kristianstad biogas (municipal waste management); Findus, Scan, Örtofta, Procordia, Örtofta Nordic Sugar (food industry/food processors); Lunds Energi, E.ON, Öresundskraft (energy companies); OKQ8, Preem, Shell (oil companies); Skånetrafiken, Veolia (public transport); WTM AB (waste collection equipment); Norups Gård, Purac, Purac Läckeby, Flinga Gårdsgas, Götene Gårdsgas (plant engineering companies); BioPreplant; MSI Teknik AB, Xylem, Spirac; Läckeby Products (pre-treatment equipment); Malmbergs Water, Terracastus Technologies, Cryo, Purac Puregas (biogas upgrading equipment); Compower (gas engines and gas turbines).
2. As the TIS framework has a weakness in explaining regional differences in technology development and evolution (Coenen *et al.*, 2012), the paper contributes by this means also to the literature on socio-technical transitions. In the majority of studies, TIS are set national boundaries that neglect local/regional specificities of TIS characteristics (for an exception, see, e.g. Binz & Truffer, 2012).
3. In particular we should acknowledge that many of the patterns found in this study may be specific for clean-tech industries and clusters.

References

Arthur, W. B. (1994) *Increasing Returns and Path Dependence in the Economy*. (Ann Arbor, MI: University of Michigan Press).

Asheim, B., Boschma, R. & Cooke, P. (2011) Constructing regional advantage: Platform policies on related variety and differentiated knowledge bases, *Regional Studies*, 45(7), pp. 893–904.

Asheim, B., Bugge, M. M., Coenen, L. & Herstad, S. (2013) *What does evolutionary economic geography bring to the policy table? Reconceptualising regional innovation systems*. CIRCLE working paper 2013/05.

Bergek, A., Jacobsson, S., Carlsson, B., Lindmark, S. & Rickne, A. (2008) Analyzing the functional dynamics of technological innovation systems: A scheme of analysis, *Research Policy*, 37(3), pp. 407–429.

Binz, C. & Truffer, B. (2012) Technological innovation systems in multiscalar space. Analyzing an emerging water recycling industry with social network analysis, *Geographica Helvetica*, 66(4), pp. 254–260.

Boschma, R. A. (1997) New industries and windows of locational opportunity: A long-term analysis of Belgium, *Erdkunde*, 51(1), pp. 12–22.

Boschma, R. A. & Frenken, K. (2006) Why is economic geography not an evolutionary science? Towards an evolutionary economic geography, *Journal of Economic Geography*, 6(3), pp. 273–302.

Boschma, R. A. & Frenken, K. (2009) Some notes on institutions in evolutionary economic geography, *Economic Geography*, 85(2), pp. 151–158.

Boschma, R. A. & Frenken, K. (2011a) Technological relatedness and regional branching, in: H. Bathelt , M. P. Feldman & D. F. Kogler (Eds) *Beyond Territory: Dynamic Geographies of Knowledge Creation, Diffusion, and Innovation*, pp. 64–81 (London: Taylor and Francis/Routledge).

Boschma, R. A. & Frenken, K. (2011b) The emerging empirics of evolutionary economic geography, *Journal of Economic Geography*, 11(2), pp. 295–307.

Boschma, R. A. & Martin, R. (2007) Constructing an evolutionary economic geography, *Journal of Economic Geography*, 7(5), pp. 537–548.

Boschma, R. A. & Martin, R. (2010) *The Handbook of Evolutionary Economic Geography* (Cheltenham: Edward Elgar).

Callon, M. (1998) Introduction: The embeddedness of economic markets in economics, *The Sociological Review*, 46(S1), pp. 1–57.

Carlsson, B. & Stankiewicz, R. (1991) On the nature, function and composition of technological systems, *Journal of Evolutionary Economics*, 1(2), pp. 93–118.

Coenen, L., Benneworth, P. & Truffer, B. (2012) Toward a spatial perspective on sustainability transitions, *Research Policy*, 41(6), pp. 968–979.

Cohen, W. M. & Levinthal, D. A. (1990) Absorptive capacity: A new perspective on learning and innovation, *Administrative Science Quarterly*, 35(1), pp. 128–153.

David, P. A. (1985) Clio and the economics of QWERTY, *The American Economic Review*, 75(2), pp. 332–337.

Dosi, G. (1982) Technological paradigms and technological trajectories: A suggested interpretation of the determinants and directions of technical change, *Research Policy*, 11(3), pp. 147–162.

Energimyndigheten [The Swedish Energy Agency]. (2012) *Produktion och användning av biogas år 2011* [Production and use of biogas in 2011].

Ericsson, K., Nikoleris, A. & Nilsson, L. J. (2013) *The Biogas Value Chain in the Swedish Region of Skåne*. Project Report. Lund: Environmental and Energy System Studies, Department of Technology and Society, Lund University. Available at http://www.topnest.no/attachments/article/12/Value%20Chain%20Analysis%20Biogas_Skane_20130828.pdf (accessed 30 April 2014).

Essletzbichler, J. (2009) Evolutionary economic geography, institutions, and political economy, *Economic Geography*, 85(2), pp. 159–165.

Frenken, K., Van Oort, F. & Verburg, T. (2007) Related variety, unrelated variety and regional economic growth, *Regional Studies*, 41(5), pp. 685–697.

Geels, F. W. (2002) Technological transitions as evolutionary reconfiguration processes: A multi-level perspective and a case-study, *Research Policy*, 31(8–9), pp. 1257–1274.

Gertler, M. (2010) Rules of the game: The place of institutions in regional economic change, *Regional Studies*, 44(1), pp. 1–15.

Grabher, G. (2009) Yet another turn? The evolutionary project in economic geography, *Economic Geography*, 85(2), pp. 119–127.

Granovetter, M. & McGuire, P. (1998) The making of an industry: Electricity in the United States, *The Sociological Review*, 46(S1), pp. 147–173.

Hekkert, M. P., Suurs, R. A. A., Negro, S. O., Kuhlmann, S. & Smiths, R. E. H. M. (2007) Functions of innovation systems: A new approach for analyzing technological change, *Technological Forecasting and Social Change*, 74(4), pp. 413–432.

Johnson, A. & Jacobsson, S. (2001) Inducement and blocking mechanisms in the development of a new industry: The case of renewable energy technology in Sweden, in: R. Coombs, K. Green, A. Richards & W. Walsh (Eds) *In Technology and the Market, Demand, Users and Innovation*, pp. 89–111 (Cheltenham: Edward Elgar).

Kemp, R. (1994) Technology and the transition to environmental sustainability—The problem of technological regime shifts, *Futures*, 26(10), pp. 1023–1046.

Länsstyrelsen i Skåne län [The County Administrative Board of Scania]. (2011) *Biogaspotential i Skåne* [Biogas potential in Scania].

MacKinnon, D., Cumbers, A. & Chapman, K. (2002) Learning, innovation and regional development: A critical appraisal of recent debates, *Progress in Human Geography*, 26(3), pp. 293–311.

MacKinnon, D., Cumbers, A., Pike, A., Birch, K. & McMaster, R. (2009) Evolution in economic geography: Institutions, political economy, and adaptation, *Economic Geography*, 85(2), pp. 129–150.

Markard, J. & Truffer, B. (2008) Technological innovation systems and the multi-level perspective: Towards an integrated framework, *Research Policy*, 37(4), pp. 596–615.

Martin, R. (2010) Rethinking regional path dependence: Beyond lock-in to evolution, *Economic Geography*, 86(1), pp. 1–27.

Martin, R. & Sunley, P. (2006) Path dependence and regional economic evolution, *Journal of Economic Geography*, 6(4), pp. 395–437.

Martin, R. & Sunley, P. (2011) Conceptualizing cluster evolution: Beyond the life cycle model?, *Regional Studies*, 45(10), pp. 1299–1318.

Menzel, M. P. & Fornahl, D. (2010) Cluster life cycles—Dimensions and rationales of cluster evolution, *Industrial and Corporate Change*, 19(1), pp. 205–238.

Morgan, K. (2012) Path dependence and the State: The policits of novelty in old industrial regions, in: P. Cooke (Ed) *Re-framing Regional Development: Evolution, Innovation, Transition. Regions and Cities*, pp. 318–340 (Abington: Routledge).

Nooteboom, B. (2000) *Learning and Innovation in Organizations and Economies* (Oxford: Oxford University Press).

Nygaard Tanner, A. (2012) *The Geography of Emerging Industry. Regional Knowledge Dynamics in the Emerging Fuel Cell Industry* (Lyngby: DTU).

Porter, M. E. (1998) Clusters and the new economics of competition, *Harvard Business Review* 76(6), pp. 77–90.

Region Skåne [Region Scania]. (2011) *Skånes färdplan för Biogas* [Scnaia's roadmap for biogas].

Rip, A. & Kemp, R. (1998) Technological change, in: S. Rayner & E. L. Malone (Eds) *Human choice and climate change, Vol 1*, pp. 327–399 (Columbus, OH: Batelle Press).

Scott, A. J. & Storper, M. (1987) High technology industry and regional development: A theoretical critique and reconstruction, *International Social Science Journal*, 39, pp. 215–232.

Shin, D. H. & Hassink, R. (2011) Cluster life cycles: The case of the shipbuilding industry cluster in South Korea, *Regional Studies*, 45(10), pp. 1387–1402.

Simmie, J. (2012) Path dependence and new technological path creation in the Danish wind power industry, *European Planning Studies*, 20(5), pp. 753–772.

Skånetrafiken. (2012) *Skånetrafikens miljöredovisning 2011* [Skånetrafiken's environmental report].

Storper, M. & Walker, R. (1989) *The Capitalist Imperative: Territory, Technology, and Industrial Growth* (New York: Wiley-Blackwell).

Strambach, S. (2010) Path dependence and path plasticity: The co-evolution of institutions and innovation—The German customized business software industry, in: R. Boschma & R. Martin (Eds) *The Handbook of Evolutionary Economic Geography*, pp. 406–431 (Northampton: Edward Elgar).

Strambach, S. & Klement, B. (2013) Exploring plasticity in the development path of the automotive industry in Baden-Württemberg: The role of combinatorial knowledge dynamics, *Zeitschrift für Wirtschaftsgeographie*, 57(1–2), pp. 67–82.

Trippl, M. & Otto, A. (2009) How to turn the fate of old industrial areas: A comparison of cluster-based renewal processes in Styria and the Saarland, *Environment and Planning A*, 41(5), pp. 1217–1233.

Uyarra, E. (2010) What is evolutionary about "regional systems of innovation"? Implications for regional policy, *Journal of Evolutionary Economics*, 20(1), pp. 115–137.

Perspectives on Cluster Evolution: Critical Review and Future Research Issues

MICHAELA TRIPPL*, MARKUS GRILLITSCH*, ARNE ISAKSEN** & TANJA SINOZIC[†]

*CIRCLE, Lund University, Lund, Sweden, **Department of Working Life and Innovation, University of Agder, Grimstad, Norway, [†]Institute for Multi-Level Governance and Development, Vienna University of Economics and Business, Vienna, Austria

ABSTRACT *The past two decades have witnessed an ever-growing scholarly interest in regional clusters. The focus of research has mainly been on exploring why clusters exist and what characteristics "functioning" clusters possess. Although the interest in more dynamic views on clusters is not new, in recent years, however, greater attention has been paid to providing better explanations of how clusters change and develop over time, giving rise to an increasing popularity of the cluster life-cycle approach. This paper discusses the key ideas and arguments put forward by the main protagonists of this approach and identifies several missing elements, such as indifference to place-specific factors, neglect of multi-scalar impacts and underappreciation of the role of human agency. Based on this critical assessment, a number of suggestions for future research are made. We argue that there is a need to study the influence of the wider regional environment on cluster evolution and to explore how cluster development paths are influenced by a multiplicity of factors and processes at various spatial scales. Finally, it is claimed that future research should pay more attention to the role of human agents and the ways they shape the long-term development of regional clusters. We outline how future studies can tackle these issues.*

1. Introduction

Over the past two decades or so, there has been an enormous scholarly and policy interest in regional clusters. While the term "cluster" was introduced by Michael Porter in the 1990s (Porter, 1990, 1998), the origins of the notion can be traced back to Marshall's (1920) influential work on industrial districts. Nowadays, clusters have become an essential ingredient of regional development and innovation strategies in many parts of the world, and notwithstanding a number of critical evaluations (see, for instance, Martin & Sunley, 2003), the notion of clusters is widely used in economic geography, innovation studies and related disciplines.

An extensive body of work has focused attention on explaining why clusters exist and what the main characteristics of "functioning" or fully developed clusters are. Compared to these static approaches, dynamic views on long-term cluster evolution have received less attention. Arguably, the question of how clusters develop and change over time is not new (Frenken *et al.*, 2014) and has been a subject of research in the past (see, for instance, Storper & Walker, 1989; Tichy, 2001). In recent years, there has been a growing recognition of the need to further develop and elaborate on dynamic perspectives on clusters to gain better insights into potential forms of their long-term evolution (see, for instance, Lorenzen, 2005; Bergman, 2008; Menzel & Fornahl, 2010; Isaksen, 2011). A popular model in this field of research is the cluster life-cycle approach. It has been heavily inspired by earlier work on (spatial) product life cycles (Vernon, 1966; Cox, 1967; Thompson, 1968; Utterback & Abernathy, 1975) and studies of industry life cycles (Klepper, 1997, 2007).

The cluster life-cycle approach has many merits. It has essentially contributed to moving beyond the overly dominant static perspectives on clusters, enhancing our understanding of the crucial factors that may trigger the rise and further development of regional clusters. Life-cycle models of cluster development, however, suffer from several shortcomings, most notably from a limited appreciation that has been given thus far to the role played by the wider regional environment, the influence of factors at higher spatial scales and the ways by which agents and their activities impact the evolution of clusters.

The main aims of this paper are to critically review the key ideas and important contributions to the cluster life-cycle approach, identify missing elements in conceptual and empirical work and suggest a set of core issues that deserve due attention in future work on long-term cluster development. It is argued in this paper that a stronger emphasis on place specificity and multi-scalarity of cluster evolution is necessary and that the role of human agents in long-term cluster development needs to be further examined. We claim that consideration of these issues might enrich our explanations of how clusters evolve and change over time.

The remainder of the paper is organized as follows. Section 2 introduces the cluster life-cycle concept, highlighting its theoretical origins and discussing its main elements and merits. In Section 3 a set of factors that have thus far been underappreciated by scholarly work on cluster life cycles is identified. Section 4 elaborates on an agenda for future research activities that may add promising further insights into how clusters develop in the longer term. Finally, Section 5 summarizes the main arguments.

2. Cluster Life-Cycle Approach

Introducing a dynamic perspective to studies on clusters, the cluster life-cycle approach has received considerable attention (for an overview, see Bergman, 2008). A central hypothesis of this approach is that clusters change with discernible phases of growth (Swann, 1998; Braunerhjelm & Feldman, 2006). Each phase is characterized by specific features influencing shifts that are assumed to be generalizable across cluster populations.

The literature on cluster life cycles builds on a long tradition in regional science and economic geography of cyclical approaches to economic development (for an overview, see Trippl, 2004; Frenken *et al.*, 2014). Well-known concepts include the spatial product life-cycle approach, the profit cycle approach, the notion of long waves of regional development and the industry life-cycle approach. These approaches derive the cyclical development of clusters from innovation patterns that follow from the life cycle of products (Vernon, 1966; Cox, 1967), evolution of new radical key technologies that underpin long waves of economic development (Booth, 1987; Marshall, 1987), industry life cycles (Audretsch & Feldman, 1996; Klepper, 1997) and the profit cycles of industries (Markusen, 1987). Also Storper and Walker's (1989) theory of geographical industrialization focuses on the link between the geography of production and the evolution of industries and technologies. These concepts suggested that the cycles are characterized by typical patterns of firm growth, firm exit and entry, international trade and locations of production and innovation. The theoretical arguments of these earlier approaches have been taken up in the current cluster life-cycle literature.

A broad distinction can be made in approaches that focus on industry-driven explanations and those that emphasize processes specific to clusters (Martin & Sunley, 2011). Industry-driven explanations relate the growth of clusters to development stages in specific industries and technologies. In general, the importance of clusters is thought to be highest in the early development phases of an industry and technology, when much experimentation takes place and knowledge is not yet codified and standardized (Ter Wal & Boschma, 2011). "Cluster-specific" cycle views (Pouder & St. John, 1996; Iammarino & McCann, 2006; Maskell & Malmberg, 2007) suggest that clusters can grow or decline independently of the development of the industry, for reasons such as homogeneity or heterogeneity in competencies, and cluster-specific technological or institutional lock-ins.

A starting point for explaining cluster emergence is that clusters take off as Marshallian economies of scale and spillovers gain momentum, increase firm profits and create favourable business conditions that facilitate spin-offs and attract new firms (Maskell & Kebir, 2006; for a critical review of the literature on the importance of localization economies, see Frenken *et al.*, 2014). Cluster emergence is partially caused by what has previously happened in the region (and globally), and thus relates to already available local capabilities, routines and institutions (Boschma & Frenken, 2011). "Even in the case of radical technological development, knowledge production is also highly cumulative and builds on pre-existing localized scientific and technological resources" (Tanner, 2011, p. 24). Other variants of the cluster life-cycle approach consider the actual location of new clusters as the result of historical accidents, suggesting that new clusters often "start out in a particular location more or less by chance" (Maskell & Malmberg, 2007, p. 612), making it difficult to predict when and where clusters will arise (Braunerhjelm & Feldman, 2006). Cluster emergence differs by industry. Clusters of high-tech industries have, for example, been traced back to discoveries generated in research, the continuous inflow of skilled graduates, as well as an entrepreneurial environment (Patton & Kenney, 2010).

The growth of clusters is characterized by a strong growth of leading firms and the entry of new firms. Firms in clusters tend to be more innovative as they have better access to still largely tacit knowledge (Maskell & Malmberg, 2007). Pouder and St. John (1996) argue that firms in clusters (or in their words "hot spots") behave differently and are more innovative because of collective learning, enhanced legitimacy and agglomeration economies. The momentum of the growth stage depends on the alignment between local conditions such as skilled labour, training, suppliers and institutions and thereby the constitution of a conducive environment for growing industries (Bergman, 2008).

Clusters enter a period of maturation or exhaustion when average firm growth, entries and exits in the cluster converge with national averages. In this phase, innovation activities are incremental, congestion costs rise, processes become standardized and easier to replicate elsewhere, and other ways of organizing production become more efficient, creating conditions for firms to leave the cluster (Swann, 1998). The number of firms falls, and existing networks become less fruitful sources of external information (Tichy, 1998, 2001). Cognitive bias, mimetic behaviour and institutional isomorphism lead to lock-ins and reduced innovativeness of cluster firms (Pouder & St. John, 1996). Different forms of lock-ins reduce the adaptability of cluster firms to changes in market and technological conditions and thereby provoke a decline of the cluster (Grabher, 1993; Hassink, 2010). Some clusters, however, are able to renew themselves by reusing certain skills and existing infrastructure and to build novel industrial and sectoral identities.

These rather generic ideas of the cluster life-cycle approach have recently been enriched by insights from the literature on evolutionary economic geography, which emphasize factors such as heterogeneity of firm competencies, the evolution of networks and a somewhat different perspective on path-dependency. Recent versions of the cluster life-cycle approach have partly adopted these factors.

Firm heterogeneity and variety feature prominently in the models developed by Menzel and Fornahl (2010) and Ter Wal and Boschma (2011). Menzel and Fornahl (2010) propose a clear distinction between cluster firms, firms in the same industry located elsewhere and firms in other industries but located in the same region. While appreciating the role of interactions between these different types of firms, the institutional context and the industry life cycle, it is argued that firm heterogeneity and localized learning processes are the central factors explaining cluster change. Menzel and Fornahl (2010) suggest that the development of technological relatedness between firms is a precondition for the emergence of a cluster while heterogeneity is considered as crucial source for the extension or renewal of development trajectories. Clusters begin "(. . .) in those regions where the knowledge bases of companies converge around technological focal points" (p. 231). Technological convergence underlying cluster formation is shaped by, among other factors, interactive learning processes between heterogeneous firms in geographic proximity to one another. Firm heterogeneity can be increased through learning with noncluster firms both locally and globally. This may bring in new knowledge to the cluster, shifting its thematic boundaries. Menzel and Fornahl's model suggests that localized learning dynamics and firm heterogeneity propel clusters through life cycles. Despite this dominant trajectory, the authors also consider alternative trajectories. For instance, without technological convergence in the emergence stage, the next stage (growth) may never be reached. Also, by introducing heterogeneity in later stages, clusters can continuously renew themselves and do not necessarily need to decline. In sum, "clusters display long-term growth if they are able to maintain their diversity" (Menzel & Fornahl, 2010, p. 218).

Ter Wal and Boschma (2011) propose a framework in which clusters co-evolve with firm capabilities, industry life cycles and networks. The authors emphasize the importance of variety as regards firm capabilities that resonates well with the model introduced by Menzel and Fornahl (2010). In addition, the framework of Ter Wal and Boschma (2011) elaborates on the effects of networks for the evolution of clusters. Ter Wal and

Boschma (2011) argue that networks and firm capabilities are interrelated so that "variety across firms in terms of capabilities drives the evolution of networks through time" (p. 923). Firms with strong capabilities will be attractive collaboration partners and thus more centrally positioned in networks. However, due to the high degree of uncertainty during cluster emergence as regards the future dominant technologies and players, networks are highly unstable. Firms are expected to switch network partners frequently, whereas the choice of partners depends both on social networks and chance events.

As clusters grow, several forces lead to stable core-periphery network patterns. This implies first that firms with superior capabilities are the most attractive network partners and thus are centrally positioned. The central network position further increases the attractiveness of these firms. Preferential attachment is stimulated by the advantageous network position of pioneers, the higher likelihood of firms in weaker positions to exit the industry, the importance of prior alliances for the formation of new ones, as well as the technological trajectory where the centrally positioned firms with superior capabilities are likely to further drive technological development. As the technology has not yet matured, tacit knowledge plays an important role, which in consequence stimulates networks in geographic proximity for reasons such as social capital, trust and the ease of face-to-face interactions. During maturity and decline, many firms exit the industry. Firms at the core of the network (and frequently located in industry clusters) tend to have a higher likelihood of survival. The endurance of networks can have, however, distinct disadvantages because of a decreasing variety of firm capabilities, which may lead to cognitive lock-in, and an increasing codification of knowledge, which reduces the need for geographic proximity. Nevertheless, a new cluster life cycle may be started if cluster firms succeed in generating a new technological breakthrough. However, similarly to the introductory stage, such technological breakthroughs will often be generated outside the cluster, leading to significant changes in the network structure (Ter Wal & Boschma, 2011).

To summarize, the cluster life-cycle approach has to be acknowledged for introducing a dynamic perspective to the frequently static literature on clusters. The approach focuses on several aspects that drive the evolution of clusters. One explanation is that clusters co-evolve with industries and technologies. However, as there is clear evidence for different development paths of clusters within specific industries, other explanations focus on cluster-specific processes and factors. In this respect, the cluster life-cycle approach has benefited from the theoretical developments in the literature on evolutionary economic geography. Accordingly, it has been suggested that the heterogeneity of firm capabilities, localized learning processes (which may reduce heterogeneity), and the openness or rigidities in firm networks are crucial factors explaining the transition between cluster phases, but also why some clusters may be able to renew themselves while others fail.

3. Limitations of the Cluster Life-Cycle Approach

This section suggests some extensions to cluster life-cycle models based on the identification of a few main shortcomings of these models. We should, however, note that the shortcomings identified below hold true only for some models but less so for others. We start out by reviewing the criticism several scholars have levelled against the rather deterministic view of cluster development in the cluster life-cycle literature. Our main concern, however, relates to the abstraction from regional context sensitivity, multi-scalar impacts and the role of human actors in theorizing on cluster dynamics.

Some authors (see, for instance, Martin & Sunley, 2011; Oinas *et al.*, 2013) have criticized the deterministic logic of the life-cycle approach to cluster development, arguing that it carries biological connotations and imply "...some sort of 'aging' process" (Martin & Sunley, 2011, p. 1300). Adopting a life-cycle approach would imply that cluster development is deemed to follow a predetermined sequence of stages from birth to growth, maturation and decline (see, for instance, the model proposed by Pouder & St. John, 1996) and possible renewal (i.e. the start of a new cycle). Also, more recent work (Maskell & Kebir, 2006; Ter Wal & Boschma, 2011) takes the prescribed sequence of phases as granted. Such a view would not allow for capturing other development patterns such as a hyper growth stage after a growth stage, or paths that never reach the subsequent stage (Bergman, 2008) and other forms of development patterns observable in the real world (Oinas *et al.*, 2013). Approaches to cluster life cycles should allow for the fact that "history matters" but also that history does not entirely restrict possible development paths. To follow evolutionary thinking then means that the development of a cluster can be explained only ex post but not ex ante since the future course of a path is open.[1]

This is in line with the view suggested by Frenken *et al.* (2014, p. 6) who state that "rather than viewing life-cycle stages as predetermined successions, the concept of a life cycle is better understood as a heuristic device to organize empirical cases into a coherent framework without denying the indeterminate outcome of the process" (Frenken *et al.*, 2014, p. 6). Adopting the latter perspective, that is, understanding the notion of a cluster life cycle as a heuristic device, implies that the criticism various scholars have raised against the deterministic logic of the life-cycle approach to cluster development is somewhat misplaced.

However, the challenge remains to explain how and why clusters can take multiple development paths. Cluster life-cycle models tend to give primary emphasis to a certain set of influences and tend to ignore others that also may affect cluster evolution. Some of the recently introduced cluster life-cycle models, for example, focus on the characteristics and dynamics of firms and their capabilities and networks, while regional characteristics, their interrelationships with extra-regional factors and the role of human agency receive little attention. In the following, we focus our attention on these often-overlooked elements in the research on cluster evolution.

Clusters operating in the same or similar industries may display different development dynamics, that is, they can reveal quite different forms of path development, influenced by region-specific factors. Saxenian's (1994) well-known comparison of clusters of computers and electronics firms in Silicon Valley and Route 128 is a prime example in this regard. Trippl and Otto (2009) have provided evidence on how regional characteristics such as the presence of research institutes, which enabled building bridges to new technologies, regional innovation culture and proactive policy approaches facilitated the revitalization of the old metal cluster in Styria (Austria), while the absence of these factors was partly responsible for the decline of the metal cluster in Saarland (Germany). Belussi and Sedita (2009) show that firms in one footwear cluster in Italy diversified their products, which attracted some big luxury brands to the district, while firms in another Italian footwear cluster carried out a cost-led strategy and outsourced production to low-cost countries. This demonstrates that no one-to-one relationship exists between industry characteristics and cluster life cycles and that, among others, regional-specific factors have not received enough attention in the life-cycle approach to cluster development. Recent conceptual work by Boschma (2014) and Isaksen and Trippl (2014) supports this view, demonstrating that regional characteristics such as the degree of diversity of industrial structures and organizational thickness as well as region-specific institutions, the presence of bridging social capital and so on are conducive to the emergence of new clusters (new regional industrial paths) and their further development.

The literature on cluster life cycles has thus far devoted relatively little attention to how cluster development paths are shaped by a multiplicity of factors at various spatial scales and their interaction. It is fair to argue that the multi-scalarity of cluster evolution remains poorly understood. For example, the relationship between various configurations of knowledge linkages at different geographical levels and cluster evolution has not been a key topic of interest in the cluster life-cycle literature, and little has been said about the insertion of clusters in institutional frameworks, which are multi-scalar in nature. Multi-scalarity is illustrated by the innovation dynamics within six globally competitive clusters in Norway analysed by Isaksen (2009). Many core firms in the clusters are incorporated in Norwegian innovation systems in the maritime and oil and gas industries. Firms find most of their partners among universities and R&D institutions inside Norway. The national system was especially important for the start and early development of the clusters as the customers on whom initial demand was based were mainly national ones, and as many firms started by commercializing the research results from large national research institutes. The growth of the clusters has, to some extent, been an integrated part of Norwegian industrial policy. The firms' value chains, however, are to a large extent global, and the firms find important innovating partners among customers and suppliers internationally. The regional level is also important as innovation processes in cluster firms utilize the unique competence, partly of a tacit nature, that is embedded in the experience of employees and in routines at the workplaces. Firms recruit, to a large extent, new staff locally, from similar firms and educational institutions that have created study programmes especially adapted to the needs of cluster firms. This study suggests that the national level was particularly important for the start of the clusters while the regional and international levels were more important when the clusters mature. A study of the nature of innovation networks in 10 regional industries found in different parts of Europe demonstrates a multi-scalar nature of knowledge sourcing (Martin, 2013). The 10 regional industries are based on different types of knowledge. Firms in all clusters source knowledge from the regional, national and international levels. However, science-based clusters are dominated by globally configured knowledge networks. Firms in engineering-based clusters exchange knowledge mainly with national and regional partners, while clusters operating in creative industries are dominated by regionalized and localized knowledge networks.

Cluster life-cycle models can also be charged as being too much occupied with the structure level, such as firm structure (heterogeneity) and knowledge networks, while agents and their activities are not sufficiently taken into account. This critique relates to Martin and Sunely (2006) who have posed the question whether path dependence does need a theory of human agency. "In what ways is path dependence intentionally created by actors, or an unintentional emergent effect at system level?" (p. 404). Maryann Feldman, in particular, has brought human agency in the form of entrepreneurship into the research of cluster evolution. Entrepreneurs are seen to "spark cluster formation and regional competitive advantage" (Feldman et al., 2005, p. 130) and also to "build resources and community" (p. 131).

The aspect of strategic action by key persons is illustrated by the re-emergence of the Antwerp diamond district after being severely reduced during the Second World War. Henn and Laureys (2010) focus in their analysis of the district's re-emergence on the role of some key persons, like two diamantaires, a representative of a diamond extraction company and a politician. Their strategic actions included securing the supply of rough diamonds to Antwerp and reducing and even stopping the supply to some competitor areas. The importance of key persons is also accentuated by Ferrucci and Porcheddu (2006) in their analysis of the emergence and growth of the ICT cluster in Sardinia, "which is a story of a few key figures (regional policy-makers, scientists, entrepreneurs, etc.) who contributed, in various ways" (p. 204). The cluster started with the establishment of an R&D institute in 1990, initiated by some regional policy-makers. The R&D institute created advanced scientific competence, acquired by a number of junior researchers. The competence was employed by a Sardinian publisher and entrepreneur who started a pioneer company in web publication. The pioneer firm was soon acquired by a large Italian company and downsized, which triggered many spin-offs by former employees. Another large Internet communication company soon arose, partly started by entrepreneurs from the pioneer firm, which spurred the establishment of further ICT firms in Sardinia.

Regional context-specific factors, multi-scalar influences as well as human agents and their activities thus do matter but have so far been underappreciated in the literature on cluster evolution. In a next step, we build on the discussion and arguments advanced above, elaborating on an agenda for future research activities on cluster life cycles that takes these factors into account.

4. Key Issues for Future Research

This section identifies some key challenges for future research activities on long-term cluster development. There are still many unresolved questions of how clusters evolve over time (see, for instance, Boschma & Fornahl, 2011). We do not intend to construct an overarching research strategy or map out a comprehensive research agenda here. This would be far beyond the scope of this paper. Instead, based on our attempt to identify missing or often-overlooked elements in the research on cluster life cycles above (Section 3), we focus on three core issues, that is, place specificity, multi-scalar influences and human agency. We show how these issues could be tackled in futures studies of cluster evolution.

4.1. *Place Specificity*

As noted in Section 3, there are strong reasons to assume that there is more than one potential development path of cluster evolution. The multiplicity of possible development paths can partly be explained by the context sensitivity of clusters. A dominant strand of the literature on cluster life cycles links the development of clusters to the development of industries and technologies while other cluster life-cycle models focus on cluster internal factors that are presumed to explain the trajectories of clusters (see also Section 2). Hence, there is some tradition in the literature that considers the sectoral and technological context. However, the regional and national contexts are also of importance for the development of industries and clusters (see, for instance, Storper, 2009) as argued in the literature on national innovation systems (Lundvall, 1992; Nelson, 1993, Edquist, 2005), varieties of capitalism (Hall & Soskice, 2001) and regional innovation systems (Cooke, 2001; Asheim & Isaksen, 2002; Tödtling & Trippl, 2005). In this subsection, we focus on the notion of regional innovation systems, which can be regarded as one of the most advanced concepts for capturing region-specific innovation-related characteristics, which are seen as important for cluster development (Trippl & Tödtling, 2008). We offer some brief thoughts on how this concept could be employed to increase our understanding of cluster evolution as a place-specific phenomenon.

The regional innovation system concept devotes attention to the companies, cluster structures, knowledge providers and the institutional set-up of a region, as well as to knowledge connections within the region and to the external world. The region is seen as a crucial level at which innovation is generated through knowledge linkages, clusters and the cross-fertilizing effects of research organizations (Asheim & Gertler, 2005). Regional innovation systems come in many shapes. There are strong reasons to assume that cluster development is linked to the configuration of regional innovation systems. Cooke (2004) distinguishes between entrepreneurial and institutional regional innovation systems and claims that the former offer excellent conditions for the development of high-tech clusters, while the latter provide a fertile ground for the evolution of traditional ones. Other scholars contrast organizationally thick regional innovation systems with thin ones and provide empirical evidence that clusters operating in the same industry but located in these different types of systems display diverging development and innovation dynamics (Chaminade, 2011; Tödtling *et al.*, 2011, 2012). Isaksen and Trippl (2014) distinguish between organizationally thick and diversified, organizationally thick and specialized and organizationally thin regional innovation systems and show through a conceptual analysis that these systems support different forms of regional industrial path development, that is, cluster development paths. These typologies are useful for analysing how the evolution of a cluster might be influenced by the regional innovation system in which it is embedded.

The relation between regional innovation systems and cluster evolution is complex. According to the regional innovation system approach, clusters form an integral part of regional innovation systems (see, for instance, Trippl & Tödtling, 2008). Empirical studies on regional innovation systems provide evidence that the emergence, growth, maturity, decline and possibly renewal of clusters are influenced by the specificities of the knowledge infrastructure, supporting organizations, institutional set-up, cultural aspects and policy actions of a particular region. For instance, regional innovation systems that already host dynamic high-tech clusters provide favourable conditions for the emergence of new ones, even if these newly emerging clusters are different from those developed earlier. Such systems offer essential conditions, such as excellent research institutes, venture capitalists, a pool of highly skilled mobile workers and dense communication networks (see, for instance, Prevezer, 2001). Regional innovation systems that are poorly endowed with such structures, experiences and knowledge assets are likely to take different routes. The rise of new (high-tech) clusters in such regions is less a spontaneous phenomenon but depends more on the inflow of external knowledge, expertise and market intelligence and a stronger role of policy (Leibovitz, 2004). In addition, new cluster formation in such regions is inextricably linked to a transformation of the regional innovation system that becomes manifest in the creation of a variety of new organizations, processes of institutional (un)learning and sociocultural shifts. Another example is the literature on the renewal of traditional clusters. Much of this work is focused on old industrial regions, emphasizing a strong relationship between the rejuvenation of mature clusters and prevailing characteristics of the regional innovation system including their transformative capacities. Conceptual and empirical research on this issue suggests that the presence or absence of favourable structures of regional innovation systems (and their changes) can make a difference, influencing the success of regeneration processes of traditional clusters (Trippl & Otto, 2009; Hassink, 2010).

Overall, there are convincing conceptual arguments and some empirical evidence that the development paths of clusters partly depend upon the architecture of regional innovation systems. Typologies proposed in the literature on regional innovation systems can help to understand why and under which conditions clusters take certain trajectories, sharpening our view of the place-specific nature of cluster evolution. Future studies may devote more attention to how the characteristics of the regional innovation system (such as the presence of other clusters or industries in the region, the quality of the knowledge infrastructure, policy actions and region-specific institutions) affect cluster evolution. However, it is not only the regional context that matters. Regional innovation systems and the clusters they host are 'open' systems, that is, they are embedded in systems at higher spatial scales. In the next section we thus examine more closely the need to adopt a multi-scalar perspective on long-term cluster development.

4.2. A Multi-scalar Perspective on Cluster Evolution

A challenge for future research on cluster evolution is to further our understanding of how factors at different geographic scales interact and influence cluster development paths. Multi-scalarity can be analysed in different ways. In this paper, we focus on two dimensions that feature prominently in the literature on economic geography, namely multi-scalar knowledge networks and institutions. We discuss why the interplay between different spatial scales as regards these two dimensions is important and how a multi-scalar perspective can help to explain cluster evolution.

The importance of different spatial scales of knowledge networks for the evolution of clusters becomes apparent in various strands of the literature. Already in the literature on innovative milieus, it has been argued that regional knowledge circulation is not sufficient but needs to be complemented by extra-regional network linkages in order to capture "external energy" for innovation processes (Camagni, 1995). Bathelt et al. (2004) claim that knowledge creation in clusters depends on the combination of local buzz and global pipelines. Local buzz describes knowledge flows through dense social networks embedded in a shared sociocultural context. Global pipelines refer to more purpose-built and formal collaborations. However, equating local and global with informal and formal networks is too simple as regional and extra-regional knowledge patterns have turned out to be more complex (see, for instance, Moodysson et al., 2008; Trippl et al., 2009; Dahlström & James, 2012; Grillitsch & Trippl, 2014). The spatial patterns of knowledge networks depend for instance on the type of region in which a cluster is embedded (Chaminade, 2011; Tödtling et al., 2012). Innovative firms in peripheral regions may compensate for a lack of knowledge available in close proximity with extra-regional networks. Furthermore, Fitjar and Rodríguez-Pose (2011) find in an analysis of Norwegian firms that a high degree of region-mindedness leads to lower degrees of innovation performance. Comparing 15 case studies in Europe, Tödtling and Grillitsch (2014) find strong sector and regional differences in knowledge sourcing patterns and innovation behaviour of firms. As clusters differ by their degree of embeddedness in multi-scalar networks, clusters are also expected to differ in their evolution (Martin & Sunley, 2007; Bergman, 2008). Clusters, understood as open, complex systems, are likely to evolve in a non-linear manner with multiple feedback mechanisms between actors within the cluster (Martin & Sunley, 2007) but also with actors outside the cluster.

As discussed earlier, Ter Wal and Boschma (2011) argue that the evolution of knowledge networks at different spatial scales is closely related to the evolution of industries and clusters. These interdependencies, which lead to typical spatial patterns of knowledge networks, may also foster lock-ins. For instance, it is more likely that networks develop in close geographic proximity and between similar firms (i.e. within a cluster). However, a too-high degree of different forms of proximity restricts the learning and innovation potential of firms (Boschma, 2005) and thereby the potential for cluster renewal and transformation. In order to avoid lock-ins, it may well be necessary to break with the typical evolution of networks along the progression of a cluster life cycle (Boschma & Frenken, 2010) and promote extra-regional networks. A crucial question for future research will therefore be how the characteristics of knowledge networks at different spatial scales affect the evolution of different types of clusters, during different phases and in different regional contexts.

Besides multi-scalar networks, the institutional environment in which cluster firms are embedded plays an important role. On the one hand, various studies have found that local traditions, values and social proximity are the defining features of a cluster's evolution (Saxenian, 1994; Sydow & Staber, 2002). On the other hand, the literature on national innovation systems (Lundvall, 1992; Nelson, 1993, Edquist, 2005) and varieties of capitalism (Hall & Soskice, 2001; Asheim & Coenen, 2006) presents compelling evidence that the national institutional framework has an important effect on the development of certain types of economic activities. In liberal market economies like the US or UK, science-driven clusters are more likely to emerge and thrive than in coordinated market economies such as Germany or the Scandinavian countries. Coordinated market economies provide more favourable conditions for sectors relying on experience-based knowledge, advanced engineering skills and interactive learning at the interface between producers and users.

Institutions of different types and spatial scales intersect in regions but we know little about "how institutional forms and the incentives they create at any one particular scale influence, are influenced by, and interact with, the institutional architectures that are erected at other geographical scales" (Gertler, 2010, p. 6). Hassink (2010) differentiates between institutional issues at the regional level comprising powerful regional actors, national-political systems allowing regional actors to influence industrial policies and supranational institutions framing the conditions for a specific industry. He studies four clusters in Germany and South Korea and argues that regional institutions alone cannot explain their development paths. Rather, one needs "to take the institutional context at all spatial levels, that is local, regional, national, and supra-national into account" (Hassink, 2010, p. 465) when analysing regional lock-ins. Martin and Sunley (2012) point to a bi-directional causality where firms and individuals influence higher-level structures like institutions erected at different spatial scales (upward causation) while at the same time higher-level structures influence the behaviour of firms and individuals (downward causation). It would be interesting to investigate how important upward and downward causation and institutions of different types and spatial scales are in specific empirical settings and specific stages of cluster development, and how this differs by types of clusters.

4.3. *The Role of Human Agency*

Investigations of the importance of upward and downward causation in specific stages of cluster evolution should take human agency into account. It is then important to realize that human agency has many forms. Individuals, (teams of) entrepreneurs and firms leaders are certainly of vital importance for the emergence, growth and transformation of clusters. However, human agency is also central for the development of higher-level structures. Policy-makers and other actors contribute, for example, to creating functioning regional innovation systems, in which global knowledge links are also often highly important (Saxenian, 2006). As alluded to in Section 4.1, mutual dependence exists between higher-level structures, such as the regional innovation system, and actors, for example, entrepreneurs, in the evolution of clusters. This is illustrated by the co-evolution of venture capital and high-tech start-ups in Israel in the 1990s (Avnimelec & Teubal, 2006). The emergence of venture capital contributed much to transforming the high-tech industry in Israel from being military dominated to becoming a start-up-intensive high-tech cluster. Co-evolution means that the appearance of a critical mass of R&D-intensive start-ups created a demand for venture capital services which spurred more start-ups and so on.

This example may also point at a general understanding of the role of human agency in cluster development. We propose to conceptualize human agency as a mid-ground between the interpretation of actors as much influenced in their decision by opportunities and restrictions provided by higher-level structures, and actors that take 'free' decisions based on their own motivation and priorities. The first interpretation resembles the path-dependent approach, which in some readings give little room for human agency. Rather human agency has 'to go with a flow of events that actors have little power to influence in real time' (Garud & Karnøe, 2001, p. 2). As a result, actors will then mostly extend existing development paths, while 'the emergence of novelty is serendipitous' (Garud & Karnøe, 2001, p. 7).

The other main approach to human agency is based on the notion of purposive action. In this view new pathways and the rise of new clusters "require social action by knowledge-able pioneering individuals, universities, companies and/or governments" (Simmie, 2012, p. 769). In a similar way is mindful deviation from existing structures by entrepreneurs said to constitute the heart of path creation. (Garud & Karnøe, 2001, p. 6). The "mindful deviating entrepreneurs" are still affected by higher-level structures in several ways. Firstly, new pathways may result from the joint contribution by a number of actors, such as economic agents, policy-makers and potential customers (Simmie, 2012). Secondly, relevant actors can create favourable framework conditions and resources through, for example, policy actions, and initiate new economic activity by mobilizing the necessary resources. Thirdly, the extent and importance of mindfulness by actors may differ from case to case and over time. Simmie (2012) describes how the wind power industry entrepreneurs in Denmark employed and gradually improved local knowledge to supply local markets in rural areas, and were then probably not deviating from existing knowledge, business models and so on. Later the emerging cluster of wind power firms was supported more strategically by particular government subsidies and tax reliefs.

It would be interesting to study the extent and importance of human agency, including which higher-level structures the agency is affected by and affects, in different types of clusters and in specific stages of cluster development. Possible approaches for this type of studies are, to some extent, illustrated by Mazzucato (2013). Her study highlights the significance of higher-level structures, but also upward causation. Mazzucato points in particular at the fundamental role of the state, through its various agencies and laboratories, in the development of radical technological innovations, and in the creation of new research-based clusters. Government labs and government-backed universities, which are often involved in large, mission-oriented research programmes in the US, have been crucial for producing radical new technologies and products.

Human agency in the form of private sector entrepreneurial activity is, however, also vital for the commercialization of new technologies. Government-funded research programmes require complementary assets in firms in order to commercialize and industrialize technological innovations. Mazzucato (2013, p. 11) documents that 'there is not a single key technology behind the iPhone that has not been State-funded'. But Apple gathered experts who were able to integrate the technologies, provide aesthetic design, perform great marketing and so on. Future studies of the role of different types of human agency in cluster evolution could learn from the approach suggested by Mazzucuto (2013). An underlying and implicit message in her study is "to follow the knowledge". Possible issues on a research agenda could include the identification of where the critical knowledge for the development of different clusters comes from, who (which persons and organization) has created the knowledge, who has put it into use, who has profited from its use and so on.

5. Summary

Over the past few years there has been a growing recognition of the need to move beyond the static views in the cluster literature and focus more attention on how regional clusters develop over time. This shift towards a stronger interest in dynamic perspectives has been accompanied by, and resulted in, a growing popularity of the cluster life-cycle approach. However, as shown in this paper, the question of how clusters evolve is not new and the application of life-cycle ideas has a long history in economic geography. The cluster life-cycle approach has many merits. It has enhanced our understanding of how clusters evolve and change and what the important driving forces of the long-term development of clusters are. We have identified several avenues for further enquiry, extending in particular to regional context-specific factors, the role of human agents and consideration of multi-scalar influences. Based on this critical assessment we elaborated on several key issues that deserve due attention in future research on cluster evolution. We focused on three main tasks. First, we claimed that one of the key challenges is to gain more insights into the place specificity of cluster evolution by examining how different types of regional innovation systems shape cluster trajectories. Second, we advanced the argument that a multi-scalar framework should be adopted to enhance our knowledge about how multiple factors at various geographical scales and their interdependencies influence cluster development. Third, we have identified the need to obtain a better understanding of the role of human agents and of how long-term cluster evolution is shaped by their activities.

Consideration of these questions in further conceptualizations and empirical investigations of cluster evolution has high potential to further increase our understanding of how regional clusters develop and change over time.

Acknowledgements

We are grateful to three reviewers and the guest editor Max-Peter Menzel for very valuable comments on previous versions of this paper. The usual disclaimer applies.

Funding

This work was supported by the European Science Foundation project on "Cluster Life Cycles" and by the Swedish Research Council.

Disclosure Statement

No potential conflict of interest was reported by the authors.

Note

1. We are thankful to an anonymous reviewer of a previous version of this paper for valuable comments that helped us to elaborate on and sharpen our argument on this issue.

References

Asheim, B. T. & Coenen, L. (2006) Contextualising regional innovation systems in a globalising learning economy: On knowledge bases and institutional frameworks, *Journal of Technology Transfer*, 31(1), pp. 163–173.

Asheim, B. & Gertler, M. S. (2005) The geography of innovation: Regional innovation systems, in: J. Fagerberg, D. Mowery & R. Nelson, R. (Eds) *The Oxford Handbook of Innovation*, pp. 291–317 (Oxford: Oxford University Press).

Asheim, B. T. & Isaksen, A. (2002) Regional innovation systems: The integration of local 'sticky' and global 'ubiquitous' knowledge, *Journal of Technology Transfer*, 27(1), pp. 77–86.

Audretsch, D. & Feldman, M. P. (1996) Innovative clusters and the industry life cycle, *Review of Industrial Organization*, 11(2), pp. 253–273.

Avnimelech, G. & Teubal, M. (2006) Creating venture capital industries that co-evolve with high tech: Insights from an extended industry life cycle perspective of the Israeli experience, *Research Policy*, 35(10), pp. 1477–1498.

Bathelt, H., Malmberg, A. & Maskell, P. (2004) Clusters and knowledge: Local buzz, global pipelines and the process of knowledge creation, *Progress in Human Geography*, 28(1), pp. 31–56.

Belussi, F. & Sedita, S. (2009) Life cycle vs. multiple path dependency in industrial districts, *European Planning Studies*, 17(4), pp. 505–528.

Bergman, E. M. (2008) Cluster life-cycles: An emerging synthesis, in: C. Karlsson (Ed.) *Handbook of Research in Cluster Theory*, pp. 114–132 (Cheltenham: Edward Elgar).

Booth, D. E. (1987) *Regional Long Waves, Uneven Growth and the Cooperative Alternative* (New York: Praeger).

Boschma, R. (2005) Proximity and innovation: A critical assessment, *Regional Studies*, 39(1), pp. 61–75.

Boschma, R. (2014) *Towards an evolutionary perspective on regional resilience.* Papers in Evolutionary Economic Geography, No. 14.09, Utrecht University.

Boschma, R. & Fornahl, D. (2011) Cluster evolution and a roadmap for future research, *Regional Studies*, 45(10), pp. 1295–1298.

Boschma, R. & Frenken, K. (2010) The spatial evolution of innovation networks. A proximity perspective, in: R. Boschma & R. Martin (Eds) *Handbook of Evolutionary Economic Geography*, pp. 120–135 (Cheltenham: Edward Elgar).

Boschma, R. & Frenken, K. (2011) Technological relatedness, related variety and economic geography, in: P. Cooke (Ed.) *Handbook of Regional Innovation and Growth*, pp. 187–197 (Cheltenham: Edward Elgar).

Braunerhjelm, P. & Feldman, M. (Eds) (2006) *Cluster Genesis. Technology-Based Industrial Development* (Oxford: Oxford University Press).

Camagni, R. (1995) The concept of innovative milieu and its relevance for public policies in European lagging regions, *Papers in Regional Science*, 74(4), pp. 317–340.

Chaminade, C. (2011) Are knowledge bases enough? A comparative study of the geography of knowledge sources in China (Great Beijing) and India (Pune), *European Planning Studies*, 19(7), pp. 1357–1373.

Cooke, P. (2001) Regional innovation systems, clusters, and the knowledge economy, *Industrial and Corporate Change*, 10(4), pp. 945–974.

Cooke, P. (2004) Integrating global knowledge flows for generative growth in Scotland: Life science as a knowledge economy exemplar, in: J. Potter (Ed.) *Inward Investment, Entrepreneurship and Knowledge Flows in Scotland—International Comparisons*, pp. 73–96 (Paris: OECD).

Cox, W. E. (1967) Product life cycles as marketing models, *The Journal of Business*, 40(4), pp. 375–384.

Dahlström, M. & James, L. (2012) Regional policies for knowledge anchoring in European regions, *European Planning Studies*, 20(11), pp. 1867–1887.

Edquist, C. (2005) Systems of innovation: Perspectives and challenges, in: J. Fagerberg, D. C. Mowery & R. R. Nelson (Eds) *The Oxford Handbook of Innovation*, pp. 181–208 (Oxford: Oxford University Press).

Feldman, M. P., Francis, J. & Bercovitz, J. (2005) Creating a cluster while building a firm: Entrepreneurs and the formation of industrial clusters, *Regional Studies*, 39(1), pp. 129–141.

Ferrucci, L. & Porcheddu, D. (2006) An emerging ICT cluster in a marginal region. The Sardinian experience, in: P. Cooke & A. Piccaluga (Eds) *Regional Development in the Knowledge Economy*, pp. 203–226 (London: Routledge).

Fitjar, R. D. & Rodríguez-Pose, A. (2011) When local interaction does not suffice: Sources of firm innovation in urban Norway, *Environment and Planning A*, 43(6), pp. 1248–1267.

Frenken, K., Cefis, E. & Stam, E. (2014) Industrial dynamics and clusters: A survey, *Regional Studies*. doi:10.1080/00343404.2014.904505

Garud, R. & Karnøe, P. (2001) Path creation as a process of mindful deviation, in: R. Garud & P. Karnøe (Eds) *Path Dependence and Creation*, pp. 1–38 (London: Lawrence Erlbaum Associates).

Gertler, M. S. (2010) Rules of the game: The place of institutions in regional economic change, *Regional Studies*, 44(1), pp. 1–15.

Grabher, G. (1993) The weakness of strong ties: The lock-in of regional development in the ruhr area, in: G. Grabher (Ed.) *The Embedded Firm: On the Socioeconomics of Industrial Networks*, pp. 255–277 (London: Routledge).

Grillitsch, M. & Trippl, M. (2014) Combining knowledge from different sources, channels and geographical scales, *European Planning Studies*, 22(11), pp. 2305–2325.

Hall, P. A. & Soskice, D. W. (2001) *Varieties of Capitalism: The Institutional Foundations of Comparative Advantage* (Oxford: Oxford University Press).

Hassink, R. (2010) Locked in decline? On the role of regional lock-ins in old industrial areas, in: R. Boschma & R. Martin (Eds) *The Handbook of Evolutionary Economic Geography*, pp. 450–468 (Cheltenham: Edward Elgar).

Henn, S. & Laureys, E. (2010) Bridging ruptures: The re-emergence of the Antwerp diamond district after World War II and the role of strategic action, in: D. Fornahl, S. Henn & M.-P. Menzel (Eds) *Emerging Clusters. Theoretical, Empirical and Political Perspectives on the Initial Stage of Cluster Evolution*, pp. 74–96 (Cheltenham: Edward Elgar).

Iammarino, S. & McCann, P. (2006) The structure and evolution of industrial clusters: Transactions, technology and knowledge spillovers, *Research Policy*, 35(7), pp. 1018–1036.

Isaksen, A. (2009) Innovation dynamics of global competitive regional clusters: The case of the Norwegian centres of expertise, *Regional Studies*, 43(9), pp. 1155–1166.

Isaksen, A. (2011) Cluster evolution, in: P. Cooke, B. Asheim, R. Boschma, R. Martin, D. Schwartz & F. Tödtling (Eds) *Handbook of Regional Innovation and Growth*, pp. 293–302 (Cheltenham: Edward Elgar).

Isaksen, A. & Trippl, M. (2014) *Regional industrial path development in different regional innovation systems: A conceptual analysis.* Papers in Innovation Studies, Paper No. 2014/17, CIRCLE, Lund University.

Klepper, S. (1997) Industry life cycles, *Industrial and Corporate Change*, 6(1), pp. 145–182.

Klepper, S. (2007) Disagreements, spinoffs, and the evolution of Detroit as the capital of the U.S. automobile industry, *Management Science* 53(4), pp. 616–631.

Leibovitz, J. (2004) Embryonic knowledge-based clusters and cities: The case of biotechnology in Scotland, *Urban Studies*, 41(5/6), pp. 1133–1155.

Lorenzen, M. (2005) Editorial: Why do clusters change? *European Urban and Regional Studies*, 12(3), pp. 203–208.

Lundvall, B.-A. (1992) *National Systems of Innovation: Towards a Theory of Innovation and Interactive Learning* (London: Pinter).

Markusen, A. (1987) *Profit Cycles, Oligopoly and Regional Development* (Cambridge: MIT Press).

Marshall, A. (1920) *Principles of Economics* (London: Macmillan).

Marshall, M. (1987) *Long Waves of Regional Development* (London: Macmillan).

Martin, R. (2013) Differentiated knowledge bases and the nature of innovation networks, *European Planning Studies*, 21(9), pp. 1418–1436.

Martin, R. & Sunley, P. (2003) Deconstructing clusters: Chaotic concept or policy panacea? *Journal of Economic Geography*, 3(1), pp. 5–35.

Martin, R. & Sunely, P. (2006) Path dependence and regional economic evolution, *Journal of Economic Geography*, 6(4), pp. 395–437.

Martin, R. & Sunley, P. (2007) Complexity thinking and evolutionary economic geography, *Journal of Economic Geography*, 7(5), pp. 573–601.

Martin, R. & Sunley, P. (2011) Conceptualizing cluster evolution: Beyond the life cycle model? *Regional Studies*, 45(10), pp. 1299–1318.

Martin, R. & Sunley, P. (2012) Forms of emergence and the evolution of economic landscapes, *Journal of Economic Behavior & Organization*, 82(2–3), pp. 338–351.

Maskell, P. & Kebir, L. (2006) The theory of the cluster—what it takes and what it implies, in: B. Asheim, P. Cooke & R. Martin (Eds) *Clusters and Regional Development*, pp. 30–49 (London: Routledge).

Maskell, P. & Malmberg, A. (2007) Myopia, knowledge development and cluster evolution, *Journal of Economic Geography*, 7(5), pp. 603–618.

Mazzucato, M. (2013) *The Entrepreneurial State. Debunking Public vs. Private Sector Myths* (London: Anthem Press).

Menzel, M.-P. & Fornahl, D. (2010) Cluster life cycles—dimensions and rationales of cluster evolution, *Industrial and Corporate Change*, 19(1), pp. 205–238.

Moodysson, J., Coenen, L. & Asheim, B. T. (2008) Explaining spatial patterns of innovation: Analytical and synthetic modes of knowledge creation in the Medicon valley life-science cluster, *Environment and Planning A*, 40(5), pp. 1040–1056.

Nelson, R. R. (Ed.) (1993) *National Innovation Systems: A Comparative Analysis* (Oxford: Oxford University Press).

Oinas, P., Höyssä, M., Teräs, J. & Vincze, Z. (2013) *How do clusters transform? In search of explanatory mechanisms*. Working paper, University of Turku.

Patton, D. & Kenney, M. (2010) The role of the university in the genesis and evolution of research-based clusters, in: D. Fornahl, S. Henn & M. P. Menzel (Eds) *Emerging Clusters: Theoretical, Empirical and Political Perspectives on the Initial Stage of Cluster Evolution*, pp. 214–238 (Cheltenham: Edward Elgar).

Porter, M. E. (1990) *The Competitive Advantage of Nations* (New York: The Free Press).

Porter, M. E. (1998) Clusters and the economics of competition, *Harvard Business Review*, November–December, pp. 77–90.

Pouder, R. & St. John, C. H. (1996) Hot spots and blind spots: Geographical clusters of firms and innovation, *Academy of Management Review*, 21(4), pp. 1192–1225.

Prevezer, M. (2001) Ingredients in the early development of the U.S. biotechnology industry, *Small Business Economics*, 17(1/2), pp. 17–29.

Saxenian, A. (1994) *Regional Advantage: Culture and Competition in Silicon Valley and Route 128* (Cambridge, MA: Harvard University Press).

Saxenian, A. (2006) *The New Argonauts: Regional Advantage in a Global Economy* (Cambridge, MA: Harvard University Press).

Simmie, J. (2012) Path dependence and new technological path creation in the Danish wind power industry, *European Planning Studies*, 20(5), pp. 753–772.

Storper, M. (2009) Roepke lecture in economic geography. Regional context and global trade, *Economic Geography*, 85(1), pp. 1–21.

Storper, M. & Walker, R. (1989) *The Capitalist Imperative. Territory, Technology, and Industrial Growth* (New York: Basil Blackwell).

Swann, G. M. P. (1998) Towards a model of clustering in high technology industries, in: G. Swann, M. Prevezer & D. Stout (Eds) *The Dynamics of Industrial Clusters*, pp. 52–76 (Oxford: Oxford University Press).

Sydow, J. & Staber, U. (2002) The institutional embeddedness of project networks: The case of content production in German television, *Regional Studies*, 36(3), pp. 215–227.

Tanner, A. N. (2011) *The place of new industries: The case of fuel cell technology and its technological relatedness to regional knowledge bases.* Papers in Evolutionary Economic Geography #11.13, Utrecht University.

Ter Wal, A. L. J. & Boschma, R. (2011) Co-evolution of firms, industries and networks in space, *Regional Studies*, 45(7), pp. 919–933.

Thompson, W. R. (1968) Internal and external factors in urban economies, in: H. S. Perloff & L. Wingo (Eds) *Issues in Urban Economics*, pp. 43–62 (Baltimore, MD: John Hopkins University Press and Resources for the Future).

Tichy, G. (1998) Clusters: Less dispensable and more risky than ever, in: M. Steiner (Ed.) *Clusters and Regional Specialisation*, pp. 226–237 (London: Pion).

Tichy, G. (2001) Regionale Kompetenzzyklen—Zur Bedeutung von Produktlebenszyklus-und Clusteransätzen im regionalen Kontext, *Zeitschrift für Wirtschaftsgeographie*, 45(3/4), pp. 181–201.

Tödtling, F. & Trippl, M. (2005) One size fits all? Towards a differentiated regional innovation policy approach, *Research Policy*, 34(8), pp. 1203–1219.

Tödtling, F., Lengauer, L. & Höglinger, C. (2011) Knowledge sourcing and innovation in 'thick' and 'thin' regional innovation systems—comparing ICT firms in two Austrian regions, *European Planning Studies*, 19(7), pp. 1245–1276.

Tödtling, F., & Grillitsch, M. (2014) Types of innovation, competencies of firms, and external knowledge sourcing—findings from selected sectors and regions of Europe. *Journal of the Knowledge Economy*, 5(2), pp. 330–356.

Tödtling, F., Grillitsch, M. & Höglinger, C. (2012) Knowledge sourcing and innovation in Austrian ICT companies—how does geography matter? *Industry and Innovation*, 19(4), pp. 327–348.

Trippl, M. (2004) *Innovative Cluster in alten Industriegebieten* (Münster: LIT).

Trippl, M. & Otto, A. (2009) How to turn the fate of old industrial areas: A comparison of cluster-based renewal processes in Styria and the Saarland, *Environment and Planning A*, 41(5), pp. 1217–1233.

Trippl, M. & Tödtling, F. (2008) Cluster renewal in old industrial regions—continuity or radical change? in: Ch. Karlsson (Ed.) *Handbook of Research on Clusters*, pp. 203–218 (Cheltenham: Edward Elgar).

Trippl, M., Tödtling, F. & Lengauer, L. (2009) Knowledge sourcing beyond buzz and pipelines: Evidence from the Vienna software sector, *Economic Geography*, 85(4), pp. 443–462.

Utterback, J. & Abernathy, W. (1975) A dynamic model of process and product innovation, *Omega*, 3(6), pp. 639–656.

Vernon, R. (1966) International investment and international trade in the product life cycle, *Quarterly Journal of Economics*, 80(2), pp. 190–207.

Index

Note: 'f' after a page number indicates a figure; 't' after a page number indicates a table.

For Product Safety Concerns and Information please contact our EU
representative GPSR@taylorandfrancis.com
Taylor & Francis Verlag GmbH, Kaufingerstraße 24, 80331 München, Germany

www.ingramcontent.com/pod-product-compliance
Ingram Content Group UK Ltd.
Pitfield, Milton Keynes, MK11 3LW, UK
UKHW051830180425
457613UK00022B/1187